DREAM HOUSE

by the same author

THE DECATUR ROAD

KENTUCKY LOVE

A FLATLAND FABLE

DREAM HOUSE

On Building a House by a Pond

JOE COOMER

Faber and Faber

BOSTON • LONDON

Published in the United States by Faber and Faber, Inc., 50 Cross Street,
Winchester, MA 01890.

Library of Congress Cataloging-in-Publication Data

Coomer, Joe.
Dream house : on building a house by a pond / Joe Coomer.
p. cm.
ISBN 0-571-12907-2 (cloth) : $19.95
1. House construction—Texas—Fort Worth Region. 2. Architecture, Queen Anne—Texas—Fort Worth Region. 3. Coomer, Joe—Homes and haunts. I. Title.
TH4809.U6C66 1991
690'.837—dc20 91-36261
CIP

Cover design by Bob Cato

Printed in the United States of America

I, on my side, require of every writer, first or last, a simple and sincere account of his own life, and not merely what he has heard of other men's lives; some such account as he would send to his kindred from a distant land . . .

HENRY DAVID THOREAU, *Walden*

ACKNOWLEDGMENTS

I would like to thank Mark Laughlin, Homer Day, Bob McDonald, Phil Coomer, Rufus Coomer and everyone else involved in the construction of our house. I would like to thank Fiona McCrae and Sarah Zaslaw of Faber and Faber and all my manuscript readers here in Texas for their help in the construction of this book. The manuscript goes into the newel post now for safekeeping.

Joe Coomer
Veal Station, Texas
July 1991

DREAM HOUSE

1

I want to be as subtle as dirt, as the accumulation of dust, or the movement of a sand dune. I have a great yearning to find blue marbles in the grass, something from the past. I want to be able to tilt my head in the way of a bird. And I am enamored of the thoroughness of striking a twelve-penny nail squarely with a sixteen-ounce hammer, of the idea of building something that will last. I would like to write fine books and keep them on shelves. I would like to be content.

2

We begin the building of our house with a marvelous discovery: kittens are absolutely appalled at being shot in the face with a water pistol. The first squirt leaves them mystified; their noses wrinkle back into their faces and they jump back with a bemused curiosity. It's the second squirt that sends them, tail high, to the far end of our garage apartment. I can't get over it.

My wife, Heather, and I have been trying for the past few weeks, in strained cooperation, to design our new house. But we haven't gotten very far; our two cats are fond of the space between our pencils and the graph paper. We've screamed, threatened and thrown them across the room, but their interest in the mouse-like pinkness and fidgeting of a pencil eraser is more powerful than their fear of broken bones. We can't draw a commode without Eliot or Blossom jumping up on the desk and taking a flying, sprung-clawed swat at the end of the pencil, sending our commode skidding into the living room or kitchen. But now we've made some progress; we're up to the second story already. It's the water pistol that has enabled this. It's something: the way their noses work back toward their ears like that, the little drops of water still hanging from their whiskers.

We're designing a big Victorian, our first house: a Victorian because we love nooks and crannies, stained glass, lots of windows, fretwork and, as the magazines say, "the warmth of wood"; a big Victorian because we now live in a small converted garage crammed to the rafters with the varied collections of a couple who run two antique malls and who, a while back, graduated with degrees in English. We've got too much stuff, haven't thrown away a book since we were able to grasp, and don't care if we do have double copies of *Old Yeller* and *Where the Red Fern Grows*, we aren't tossing any of them out. It might stir up those old tears.

Size and style seem to be the only things we agree upon in a house. Strained cooperation. It seems to me that two people who love each other as much as Heather and I do might be able to form an agreement on, say, the pitch of a roof or the number of sinks in a master bath. I'll work at my desk for hours at a time, water pistol ready, designing my heart out, and she'll breeze in, place a tiny but severely sharp finger on my drawing and say, "That won't work." The cats get up and leave the room of their own accord when they hear Heather say this.

"But," I say, "it's the only practical place for it."

"Haven't you ever heard of a work triangle, Spud?" she asks.

(I'll let you in on a secret: she knows I don't like to be called Spud. We've never specifically discussed this, but she's my wife and she knows it. The word spud has no relation to me. I'm a grown man.)

On the other hand, I feel it is my duty to let her know when she puts a bathroom under a stairway. "I won't be able to stand up and pee," I say. "I'll bump my head."

She looks up and says, "It will make you understand women better. Relax and sit down."

And so it goes. We stare each other down; we erase or make our lines darker; she doesn't know my name for two days; a shadow is upon both our houses; the cats get diarrhea—all of this over the size of the windows above the sink. Our parents urge "the spirit of compromise in marriage," and Heather and I both gurgle insults and remind them of old, uncalled-for spankings.

It's not that we don't see our problem. Heather is after beauty, utility, ease of movement, airiness, a big, happy home in which to raise a family. And I, at times, refuse to discuss it.

I set back to work, the halo from my desk lamp surrounding me, my graph paper and ruler. I work on the library, my own library, an English major's lifelong dream, first measuring books, then designing shelves for them, drawing in a smooth oak floor that I can whiz across in my wheeled desk chair. I put a huge, stuffed moose head opposite the stained-glass windows. Yes. I wonder if I can get Thoreau or Conrad's head done in stained glass. Suddenly, I smell pipe tobacco. There is a beautiful novel being written. The final epiphany washes warmly over me.

And Heather places a finger on my moose head and says, "That won't work."

I mumble, stunned, forgetting sentences, the spell broken, Conrad falling down in shards upon my head.

"That won't work."

"You said I could do the library my way," I plead.

"As long as you left room for the laundry chute. Right here," and she points, "where this thing labeled 'moose' is."

I am sure, I mean I am absolutely positive that our house will be built. I have no doubt of it—we will look back on this time with nothing but good memories. Heather, my wife, is the most beautiful, intelligent person I've ever met. I fall asleep thinking of her and my life is graced by her presence.

I stand up from the desk, pick up the loaded water pistol. She steps back and pulls from her jacket a water bazooka, Motts Five and Dime written all over it. The bazooka looks like it will hold two quarts of water.

"Anarchist," she yells.

"Fragrance and love," I whisper.

Three feet apart, point-blank range, we empty our weapons.

3

I wake up every morning appalled that things have changed, appalled at the accretion of sediment. It's still hard to believe at midday, and by nightfall I've got to say something.

4

In the late fall of 1985, when Heather and I decided where on my parents' farm we'd build our house, we were careful to lay our jackets on the leaf-moist ground of our site and make love there. The branches above were bare, the wind cold to pink, and so we had to hurry.

5

At the age of eight or nine I once told an unfamiliar woman I was born in the belly of a B-36 bomber. My father was in the Air Force when I was born, a tailgunner in a B-36, and this birth lie sounded reasonable even to me as I uttered it. What's remarkable is, the woman believed. A life born of a lie. I lied and smiled as sweetly as a bird on barbed wire. I almost reached out and took her hand.

6

For some reason, we had a great deal of faith in a rock wall. So that's what we did first. I think it had something to do with Heather's New England heritage and my love of building forts. The last white man killed by an Indian in Parker County, Texas, was on his way from our farm to Weatherford. This was in 1878, but we've found three arrowheads on the farm lately, and who can tell if they were shot two hundred years ago or barely missed my scalp last week? We felt we would sleep better with a rock wall surrounding our home. We wanted to establish our territory and watch moss grow over the span of our lives. We envisioned sitting on our rock wall in the sun, holding the hot breath of a horse's nose in our palms.

I think the most memorable characters in books, and in life, are those who fight something so big it can't be overcome, people who in a way rail against any sure thing, like the past, or God, or rocks. In the building of that rock wall I somehow keep more than pain. The skin, nail, hair and bone lost is of little consequence. What more I keep, and it is far more painful than the memory of the physical pain, is the understanding that my ancestors were better men than I. I had always known that my father and grandfathers were better than me, smarter, stronger, knowledgeable on a far wider scale, but I still had the idea that my great-grandfathers and their fathers must have been a bit inferior: primitive ancestors, old dolts. I don't know why I held this notion.

I found, at the age of twenty-seven, college degree and fairly keen fastball, that I could not balance one rock upon another. My ego, at an all-time high because I was about to marry the girl I wanted, was, to use a phrase, between a rock and a hard place. For hours at a time I'd put one rock upon another to watch it tumble down. I'd gather a great many rocks together at my feet and test-fit them, searching for two flat surfaces, or a concave and a convex, something with a good dovetail joint. The beginning of a rock wall is simple to the point of treachery: lay a rock on the ground. Many rocks can be found in their natural state in this form. It is the next step that confounds. And so, after many hours, I started my rock wall up against an oak tree. I could lean my second rock against the trunk. By the end of the wall I would be a criminal, one lie covering the last.

For a time, rationalizing, I was convinced it was the nature of my rocks. I was given a world unable to be worked. "A man could stack live porcupines as well as these rocks," I told Heather. She kept pushing little rocks under my big ones, shoring up my life. "You can see through all of my work," I said.

"What do you want—brick?" she said. "It's a rock wall," she said. "It's perfect in its irregularity."

I was holding a rock three times the size of her head. For a moment a cairn seemed a far more noble and romantic idea than a wall.

I know my ancestors could build rock walls. They're all over the place. Once, in the city of Bath, England, I went into an Edwardian facade, descended stairs and was confronted with a Roman rock wall. There is no forgetting, no avoiding. Every fence line, every foundation screams my inferiority. The pyramids don't boggle my imagination because my imagination can't even rise above the level of the third course of stones. The most intimidating force in my life isn't heredity or environment. It's gravity.

In the end, we attached the front-end loader to my father's tractor and brought loads of dirt from the dry pond below our house site to support one side of our rock wall. We put a sixty-degree curve in the wall so that it angled back toward another oak tree. In this way the complete wall, twelve feet long by two and a half feet tall, is contained on three sides. At most it can fall in only one direction. The curve also makes it convenient to say that we meant the wall as a sort of amphitheater, a backdrop, someday, for our children's plays. I wouldn't be surprised, I mean I wouldn't begrudge them, if they sometimes used it as a fort to fight off Indians.

7

Since I was born, when I stepped into a room with my fly lowered, every female in the family yelled, "Uncle Elmer." Raised a lip, zipped up my forgotten fly. Years went by. There I was sitting in the kitchen at Grandma's, when my Aunt Melody yelled, "Uncle Elmer"; my brother and I checked ourselves, but she was talking to a man there in the doorway. He asked Phil and me did we remember him.

"Of course," we said. But we didn't. He looked like Grandma.

"This is Grandma's brother," Aunt Melody said.

I looked at him. His fly was open.

"Uncle Elmer!" I yelled, and he took it that I really had finally remembered him, and hugged me, and I smiled the smile of knowing secrets, of growing up.

I saw Uncle Elmer only three times in my life, all in the old house on Refugio, and he's gone now, but he never disappointed us, always showed up as himself, and my children someday will occasionally go by his namesake.

8

I know what it is to want water. I've never killed anybody for a drink, but I've knocked another kid away from a fountain before because I wouldn't wait any longer. Here, where we've chosen, Texas, there's a certain scarcity. The small pond below our house site is almost always a bowl of dirt, our creek is an arroyo, and rain comes in the spring and fall, rarely between. And so, before we snap our first string line, we drill a well. Drill down 210 feet to aquifer, the Paluxi, and come up with vibrant water. It is clear, chilling as blood and broken glass, and as endless, it seems, as my arm to the pump. All things are possible after this. Thoreau himself built next to a ready source of water.

Our well is seventy-five feet up the slope from the pond, and immediately after the first gurgle and swell from the pump we think of our dry pond brimming with Paluxi: trout and pike arcing over a moonstruck surface, a gentle trickle over the spillway, croak of frog, splash of feet, a Coomer cannonball. Think of this, but close the valve. It's possible that it won't last. The town of Poolville, a few miles up the road from us, lost its namesake years back. Farming choked the hole with silt; the Indians and first settlers who used it as a landmark will never be able to find their way back home. And they say our aquifers won't last either, that they're going dry. We filled in a dangerous hand-dug well when Dad first bought the farm. It was eighty feet deep and dry as the driftwood in a cactus garden.

"Hurry, Heather," I say, as I jump back from the spigot, "get a drink." I don't know what I'd do if someone told me my water won't last: fill a few jars for the basement, look to the sky with longing, cut back on watering the yard by half. We haven't yet let the well course its way to the dry pond below.

I should be two or three feet underwater by now. I don't know how long this pond has been here, but there's a ten-inch tree growing on the dam. I don't even know if I can call it a pond. Water's been here before but it was just passing through. It came down from the slopes above (on one of which we'll build our house), around and over the roots of our thickest, stumpiest oak, made a little ravine and dropped into this bowl of dirt, where it lingered only long enough to make me think of the word water. An elephant couldn't dig it out now. When we bought the farm, came upon this hole in the ground, my first thoughts were meteorite strike, much less bluegill, let alone skip of stone. Perhaps it's because the pond is man-made. Texas has few natural bodies of water. I think all of its lakes and ponds look a little embarrassed, dams across their bottoms, people driving across them. There are five ponds on the farm. Only one holds water consistently.

We have seen water in our pond. The morning after an all-night spring thunder we walked down through the oaks and found the pond actually full. So we know we have a fine watershed. But the next day the pond was its old self again: water so clear you could see ants crawling on the bottom. The thin water of our pond rivals Walden in its crystal clarity. Where the water went is beyond me. It didn't flow over, and we've found no appreciable leach hole. The fields below the pond show no fantastic erosion. We may take Thoreau's idea and throw a bottle of red dye into the pond after some deluge, then wait for a newspaper story chronicling the event of a landlocked red tide, telling us on which doorstep our pond ended up. It left too quickly to have evaporated, so I don't think we'd be rained on red. I suppose it's possible the water finds its way back to the Paluxi, and we'd have Kool-Aid for our bath and breakfast.

But now, the pond in winter is silt-ridden, brambles on the dam, brush on the shore, more leaves than water in the basin. The sky has only this reflection. There are gopher mounds in the shallows, a few deer tracks along the bank, and our dogs have left their ways in one or two places. Thoreau surveyed and plumbed Walden in the winter. He could walk across the surface on the foot-thick ice. His pond and mine are different, it's true, but I can walk across this pond in winter too, even though it's in Texas and this day is a mild forty-four degrees. He found what he thought might be a universality in the length, breadth and depth of his pond, and I have no room to differ. The very depth and intensity of our lives are always at that point where the breadth of our past and span of our future cross, this day in the bottom of a dry pond, whether we'll tell the truth next we speak. I like my empty pond. The water will come, and go. It has no fish but it has potential. Some man made it. It does something as honest and patient as waiting for the rain. It has no reflection, but is its own true self. On our farm, in our lives, it has been and always will be a landmark.

I have a great deal of fun running down one bank and up the other.

9

My memory is an insincere and common one. I've always known this. I think I would rather break my toe on a cobblestone than try to remember where it is. My wife is appalled that I have no memories of my life before the age of five. She claims she can remember her mother's scream. I know my memory is faulty, and guilty, from my re-reading of my old journals. I am consistently surprised, aghast, ashamed and, most frequently, unbelieving: surprised that I wrote so badly, aghast that I was so naive, ashamed that I acted that way in the first place and, finally, rationally, not believing that what I hold in my hands is my own journal. I don't even recognize myself. I wonder how my mother knows me. It must be by smell. If I were my dog, I'd bite me. I'm trustworthy only in the present tense.

10

In winter the deer come down through the brush of our watershed and cross, among the oaks, to the fields below. I know people must have followed them, meat on the hoof. My brother killed a nine-point buck two seasons ago on the trail. Except for the shell cartridge and the Elmer Fudd cap, he could have been any hunter of the past twenty thousand years.

We weren't the first to choose this place, not by a long shot. The arrowheads we've found are evidence of that. And there've been other finds too: old buggy parts, a brass lock, a heart-shaped child's stirrup, a depression in the soil, a brick in the creek. Our farm is at the headwaters of Ash and Woody creeks and, apparently, it's always been a chosen place to live.

This is my father's farm, his land, and he has, therefore, the best eyes for finding things in his dirt. The dart point he found not fifty yards from our house site is called a Yarbrough by the reference books. It's an archaic point, prehistory, made most likely between 500 B.C. and 1000 A.D. This is a great deal of time to make an arrowhead in, but I suppose it takes practice to get it right: rock against rock. The point washed off the hillside and lay exposed in our dirt road. I don't know how it came to be on our hill, whether it was a bad shot lost in the grass, or was buried in the flank of a bison who ran off with it and later died on our hill, or was simply shot as far as it would fly in some fit of archaic disgust. But it's here now, dark grey, still sharp and, in my palm, almost unbelievable. I can't believe it was thrown away. It must have been a prized possession then, the time and work involved, its power over life, food on the spit. It's even possible that this point was an instrument of war, two men, two tribes, fighting over our homesite. There's not much that separates us, only time, a bit of stone. I've come to know the ground under these oaks, and I think I'd fight for it too. Unless, of course, you started shooting arrows at me. Man's basic problem is that he was given the earth as a homesite, but then had to share. My little brother, Deerstalker, wanted our site, but I was firstborn, and he relinquished with just a hint of antipathy, took a site with a better view, fewer trees and much rockier ground. If he'd intended farming he might have dropped me in my tracks, the shaft quivering as I fell.

The people of the dart point were, as best I can tell, the Tonkawas, whose name for themselves was "Tickanwatick," or "the most human of people."[1] I am fond of this pride, this arrogance. What could this have meant to them? They were not, apparently, as fierce as the Apaches to the west and north, yet were fierce enough to be their neighbors. They did not have the ingenious survival techniques of the Coahuiltecans to the south, who gathered a "Second Harvest: whole seeds and similar items picked out of human feces, cooked and chewed,"[2] yet they did succumb

1. W. W. Newcomb, Jr., *The Indians of Texas* (Austin: University of Texas Press, 1986), 134

2. T. R. Fehrenbach, *Lone Star* (New York: Collier Books, 1968), 14

to the gathering of fruits and berries when the bison were few, and the foot or any other portion of a human enemy was occasionally on their menu. I don't know what human means to me yet. But I'm sure they thought they were chosen: "We the people . . . " As I'm sure I'm chosen, the most human of humans, the most misunderstood of those without understanding. I know they dipped their dart heads in mistletoe juice, for poison or for luck. I hold my point up to the light, give it a sniff. The women of the people decorated their breasts with concentric circles, radiating from the nipple. When a Tonkawa died he was laid in the center of a teepee and his mourners formed circle after circle around his body; if the dead had been killed in battle, his mother, in mourning, would slice off her nipple and never mention her son's name again. During the nineteenth century, when the Tonkawas were dying off rapidly, they borrowed Comanche and Anglo names because their stock of traditional Tonkawa names was so depleted. What human means to me. I know they liked this country. I know they own it no longer. I know they're among the dead, this stone point turned tombstone.

Captain Veal. I repeat the name, trying to understand him from the sound of my voice over the consonants. He is the skeleton in our closet before we've even built our house. The plat of our farm shows that one quarter of the original town of Veal Station, his town, lies within its boundaries. You'd never know it, driving by.

The first Anglo settled here in 1852, three years after the first person camped out at Fort Worth, twenty-eight miles to the southeast. Bill Woody picked a likely ridge, with good water below. Any farther west and the land was untenable; the trees ended, end, here, only grass beyond. I might have stopped here myself. I find great solace in trees. I'm sure it's something innate, something remembered, my arboreal past, safety in the trees. I think it would have taken great courage to confront a Tonkawa without a tree to hide behind.

The place was first known as Cream Level, the soil around Woody Creek, rich as cream. Reverend (later Captain) Veal came and opened a store, more settlers arrived, soldiers to protect them, and in 1858 the town was officially organized. The Fort Worth to Abilene, Kansas, stage made a stop here, the dynamic Reverend

Veal drew crowds to his sermons and a school was opened. The lumber to build it was brought from east Texas, a wagon journey of two hundred miles, and at the time the two-story structure with bell tower was the largest building west of Dallas for hundreds of miles. The post office opened in 1867 followed by a college in 1877. So, everything was going along just fine when Reverend Veal was assassinated at the Confederate reunion in Dallas in 1892. That seems to have taken the heart out of the community. Not a building survives. Veal's assassin, an outraged husband, was acquitted under a Texas law allowing the avenging of insults to women. It wasn't only the Indians raping on the frontier.

Heather and I have taken a metal detector and a spade up to the town site on the hill, have found the scrap of the blacksmith, and horseshoes like so many soda can pull tabs, a few bricks, a hearthstone. Across the street there's a granite marker remembering the college. It was the last building to go, burning to the ground in 1941, victim of arsonists. Its bell tower could not sound the alarm as it had for fires and Indian raids before, because someone had stolen the big brass bell in 1917, when the metal was precious. Outrage upon outrage.

The passing of this town seems more unbelievable to me, more preposterous, than the fall of Rome. In the end it wasn't Veal that closed the town at all: shame passes; but so did the new hot top road from Weatherford. It went to nearby Springtown, and so did the post office. The school's last class was that of 1911, and then everybody moved to Springtown to get the mail.

If you climb up on top of the granite marker put up by the state, if you climb up on top of it in the winter, you can still see down the hill to the trees along Woody Creek, and to the bare oaks of our house site. We'll build on the legacy of their fallen leaves.

11

I read a book once that said I (generically I, as a writer) began to write because I thought I might die someday. This is not so. I began to write in the same way that I was born—without even thinking about it. For a time I wrote to impress girls. But now I do it to place myself in the world, in a season, in a house, in a memory. I want to remember it all.

12

We begin the rebuilding of Veal Station by cutting the tree limbs out of what will be our house. We only cut them out of the first story because that's as high as we can reach; we'll worry about the second story when we can stand on it. The trees we build among are stately for this area, pin oaks forty feet tall—though you wouldn't be able to get a six-foot piece of lumber from them. That's why Veal sent to east Texas for pine. Our trees grow slowly, over many seasons, and writhe and twist and blow in their growing so that, almost inexplicably, a twenty-inch diameter tree has a seventy-degree bend in its trunk. You'd swear God's been here, absentmindedly leaning on trees.

We place a few stakes among the oaks, marking the corners of our house. This is great fun, takes very little effort and can therefore be done with great bravado. I pound the one-by-two-inch stake with my ten-pound Thor sledge! Then we tie all the stakes together with a string and see which limbs hang over into our dining room, living room and so on. Taking my father's battered chain saw to a limb almost makes me wince: one limb fewer to jump for if a bear should suddenly appear. The bark and sawdust spiral to the leaves below. Freshly sawn oak has a stink all its own. Even my grandfather, who is the subtlest man I know, says "putrid" when he works with it. He was a fireman most of his working life and the only other time he uses the word putrid is to describe the smell of a burning human. I cut as close to the house as I can, hoping the limbs will turn up and away over the years.

I cut, the limbs and last clinging leaves fall to the ground and Heather, in her element, drags them to clear ground. The sky is grey-blue above, and Heather is blue shirt, blue eyes below, with a brand-new pair of white gloves. Some of the limbs weigh half as much as she does. I teach her to pull from the big end of the branch. She grabs hold, digs in and pulls downhill to her bleak pile, which to her disgust won't hold a match.

"How can we build a house if we can't even burn a bunch of tree limbs?" she screams.

I shut down the chain saw.

"Give it to the spring," I say. "It's too green to burn."

"Rabbits will have moved in by then," she says, and she jumps in our little truck, runs up to Mom's house and returns with a can of lighter fluid and the Sunday paper. She has a great deal of perseverance. Her favorite thing is an open world and the smoke off a fire.

We stand over the burning branches, coaxing and stomping, working toward the center and the bed of coals. The smoke moves around us, through us, up into the clearing of our house that is more open now, turning from white to grey to blue.

Build this house as you would a tree, working every day with a complete lack of faith but with the knowledge that your wife expects you to succeed completely. Your joints will come very close this way, close enough for putty to cover and for her to overlook. With luck the wood will swell and close the joint by itself. Your house will live on its own when you're finished, live and stand after you die, bear leaf, fruit, root and happily die itself someday, fall in upon itself, open up to the sky and world, clouds across above, and molder, rot, burn, birth a tree, screams of children swinging from its branches. No one will know of your joints, if they became wider or thinner as time allowed, and your wife will love you to your grave and after. She will always remember your name and how you stood on a stump, on a stone, on a ladder, reaching up, down.

We kick the last stubs into the hot coals and spend a few moments looking for other things to burn before we begin. Our faces are flushed and we look at one another, our singed eyebrows rising, and realize the cold coming. There's nothing left to do now but build a house. But the day is done, and so we turn and make our last dash across the clearing in our oaks, racing through the living room, down the hallway and out the back door.

I want a house of pride, a house of refuge, a house with vast porches where wind is not lacking, a house that you can stand in the middle of and still see earth and sky in all four directions. In the spring I want it to breathe and in the winter I want it to settle in upon itself like a big dog to his blanket, but for all seasons I want evidence of wind: scratching branches, leaves against windowpanes, scattered groans and pops of timbers. And inside also; ebbing dust motes and stillness enough to hear the world out. I wonder how music will sound there, how sound will roll around corners and down hallways, if a yell will reach from one end of the house to the other and get an answer. I wonder how smells will rise and fall, bacon from below, musk of clothes from above. Tell me if I'll be fond of my home, call it a home, secure in my home, tin and board.

13

This is the only book I've ever written without any malice aforethought—no plot, subplot, levels of meaning. I don't know what brings me here to this paragraph, or what I'll say next. I do have some notion that all the houses I have lived in and the lives I've led in them have something to do with this house, this life. I've come to this room from somewhere else. The present is so choked with the past that I can't do anything without pause. No moment comes to me so fresh that dust isn't already settling on it. What's just happened is always more intriguing to me than what's about to. The present sometimes doesn't even seem to exist at all: there's only the past and what the past allows. Many houses will become my house.

14

We stop almost before we begin. We were brilliant on paper, but on a wet, cold December 28 we put our first line level on our house site and find one end lacking six feet of earth. We've known there's a slope, but our hearts are still good, so we never expected a six-foot drop on a fifty-foot run. We check our measurements twice, bang the line level on a rock and check again. It's unbelievable. The front porch of our house will be over my head. This is not a porch to welcome friends to.

"I thought we were in Texas," Heather says.

"We might as well build a tree house," I say.

And we both sit on a cold stone atop our rock wall.

The next day we get two bids on our foundation: $11,500 and $10,900. This is a bit beyond the $3,000 we had budgeted for it. One of the contractors asks, "Well, why did you pick this hilly place with so many flat ones to be had?" It's an awful piece of irony, trying to make a flat place in Texas. So we sit in our garage apartment for three days and speak of its thrifty convenience.

"Solid little cottage," I say, pounding the walls with my fist.

"The toilet has, does and forever will leak," Heather responds, in a foundation funk. "Our children will be born, raised and eventually die in this garage apartment."

We eat nothing but Doritos and bean dip for these three days.

And then we stick our hands in our pockets and make the decision to just build the thing ourselves, foundation, house and all.

"We're young," Heather encourages.

"I built this garage apartment, didn't I?"

"Yes," she acknowledges, "but we don't have any other choice. We don't have enough money to do it any other way."

I imagine Hindenburg stuck his hands in his pockets when he decided to take the zeppelin up for one more flight. I wish we could make the decision with our dukes up, or at least while mooning the world. There would be some arrogance, some pride, in that. But I still place our decision with that 150 years ago to head west. They too, poor, desperate, must have struck out with hands in pockets, even if just because of the cold.

Perhaps we are not so brave. There's something I've been keeping from you: my father owns a lumberyard. There is nothing like the courage a fellow has when he's sitting on his dad's shoulders. So although we know we don't have enough money to build a house, we decide we can't pass it up at wholesale. My father tells us this only makes sense when you're in your twenties. But he shrugs his shoulders and offers, "Even if you lose everything, you can just start over," and he lends me his pick and shovel. I promise to return them sharper than I received them. This strikes me as pompous after I've uttered it, but when Thoreau said it of an ax it sounded fine.

Heather says, "Let me carry the shovel. I can't lift the pick."

So I take the pick, put it over my shoulder and we yodel, "Hi ho, hi ho, it's off to work we go," in a bread and butter innocence, the vast, cruel and barren stretches of the world before us.

There are probably very few things more basic to most creatures than digging a hole. But this species, and this member in particular, is not the burrowing sort. We have always preferred climbing a tree or lifting a stick to getting dirt under our nails. When we've had nothing to build and live with but dirt we've usually cut sod out of the earth and built a house on top of the ground rather than live in the hole we've just dug. But in order for our house to eventually tower some fifty-three feet above the ground, we have to begin by digging three feet down to a pumpkin-colored clay. This sounds like a more than even trade, and it's true, the pumpkin clay is a rich reward after so much black loam, but for my money a sixteen-dollar pick is a high price for the labor it enables. Why does the earth hold my pick when I strike it? Strain of pulling back, over, striking, ache of heart, lung, lung, lung. What life is this? Why does this dirt cling to this blade? I want the earth to break away like Styrofoam. I want to work with one hand, to breathe like an engine, to carve the world without effort, to strike without hope of redemption and finally to lose the fear that each stroke might be my last, through weariness or collapse of will. I don't know how people live as long as they do, how the species came this far, the way I swing a pick. Six feet down the ditch, 180 to go, my face pumpkin-colored, I bury my hatchet and decide on mechanization.

The rental of a trencher for a one-day period in the state of Texas, county of Parker, year of 1987, month of January, costs $145. This is a bargain beyond a man's wildest dreams. The breasts of women mean nothing to a man on his rented trencher. I realize instantly the pick is an instrument of chaos, crude tool of prehistoric man best used for cracking bones for marrow. A trencher is a gas-powered chain with twenty picks on it. Simple enough, but this self-propelled wonder comes with an armchair: a pick with a seat and all I have to do is steer. I was born exactly on time.

I move down my string line along the perimeter of our house digging what is known to the trade as a "footer." We plan to form up this rough ditch in a staggered way, dropping with the slope, then fill it with concrete and use Hadite block to build a level perimeter foundation. Then we'll fill this huge box, six feet deep at one end and one foot at the other, with dirt and pour a concrete slab on top of it all. If a house could be built without our figuring out what to do next it would be done in half the time.

Why must the floor be level? Who started such a fashion? I think of these things after I finish with the trencher, look over my work and see that all of the dirt has fallen back into the ditch. Heather, shovel in hand, has been cursing my wake. We spend the next four weeks shoveling dirt and building 420 feet of forms for the footer. Heather weighs in at ninety-five pounds, and when she fills a shovel and throws the dirt she usually follows it for three or four feet. It is very hard at times to tell who is shoveling whom. A great soggy clod fails to separate from her shovel when she pitches it, and the shovel and clod, with the force of an electric shock, throw her six feet away, face-down in the ditch. My wife shames the clod verbally for many minutes and finally kicks it in the side. She may be light but she outweighs most hawks and eagles.

I grin into my shirt sleeve and pound my wooden stakes into the vampire earth. Somehow this is very satisfying work. It has something of the feel of sport to it. There are the occasional glancing blows to the shin with the sledge hammer, but these are sport too. To split a pine stake truly and cleanly and have the sledge break the thin skin above the ankle and make the bone vibrate and sing is the pitch of life; the only thing better and truer is the short, circular journey that follows, away and back to the hammer to tell it you love it. In days I learn to wear thicker, higher socks and even to kick my own leg from under me to avoid striking myself. As far as I know I am the first person ever to be able to do this.

In the midst of our work the pause of winter. Snow comes to Texas like a Volkswagen in the mailbox. All we can do is cross our arms and look at each other. Even our dogs don't know what to do. We take Puppy, our black Labrador, and Butch, a big Australian and German shepherd cross, out to the farm and they lift their paws from the snow as if they've stepped in dog doo. They are Texas dogs and can't help themselves. It is their first snow of consequence, the world without not what it was. But they are soon on the trail of their own breath, sliding, snapping snow, retrieving one of my stakes. They are intrigued with the runs of my ditching, following the outline of our house, nose to ground, sure some great rodent has done this awesome work. The vex of their young lives is upon them: neither end of trench ends in burrow, nor is there rabbit between. They whine, retrace and finally look to us for answers, but we have none to offer save snowballs that need catching and the dogs bark, Yes, yes, the answer is a snowball to catch, to disappear the moment it's caught, and the answer to snow is water and to water a chase down the hill and into the pond.

It seems that even as they slide to the snowy bottom we are in the trenches again. Although snow lasts longer than rain, it rarely lasts here. But mud remains and remains. Before the first stone is laid I see we will build several houses while this house is built. The snow melts and a heavy rain afterward fills our trenches with silt. There is a great deal of beginning to beginning. Already I yearn to burn my forming stakes in our fireplace. We aren't even above ground yet and I want a plume of smoke from the chimney.

But even with shovel in hand, the day begun, sweat, ache of

shoulder, I have some notion that the present is only temporary. I won't have to move earth from now till the end of my life. Tomorrow I will be working with the lightness of spruce. I hear the wake of my ship coming in, and the Nobel committee rubs sticks together in my ears. It's not that I'm opposed to digging on a moral basis, but I think if I couldn't see an end to this sport I might run away with my wife's money.

I place a great deal of emphasis on will, the will or audacity to begin, the will to continue and sometimes the will to stop rather than finish. We're facing a project at least a year long. The only thing I've ever taken on that required this kind of effort and time was a novel, and I didn't have to do that when my fingers were cold. I was wretched at times but never runny-nosed. I never once had to hit myself in the shin with a hammer to close a chapter. It is Heather and me against the weather, the price of spruce, the lay of the land and our own speculation, the dream of a big Queen Anne Victorian bracketed in oaks. What enables us to continue with this ache and tyranny of mud isn't the present day's accomplishment, not even the week's or month's, but our projection of a life in the future, iced tea on the verandah at the end of the ditch. We spend our days moving dirt from one place to another, counting off each shovelful. Even if that last shovel of earth was a mistake, dropping back into the forms, that's one more mistake done with and we can move on to the next mistake. Time is a substance like dirt, to be moved from one place to another. Building this house is something we've been forced into by lack of funds. It's a summer job, and I think the only way we'll get through the summer is to constantly remember that fall will come again.

I spend the last half of a Friday with my father's tractor and a post hole digger, drilling pier holes in our footer to give it a better grip on the planet. The next morning, Heather, Dad and I with hoes and shovels in hand meet the first of two concrete trucks. I am sure my forms will collapse under the weight of the concrete, or, worse yet, the driver will look at them and burst out laughing. But neither event occurs. The concrete comes fluid and fast and it is all the three of us can do to spread, tamp and smooth it. Before we can even ponder, it is over. We stand up, drop our hoes and trowels and watch the concrete truck roll away. Then it happens. We turn back to our footer, level and straight and yet glistening, and we are struck with pride. We touch each other, walking around and looking at the thing from every angle. I am twenty-eight years old but I look to my Dad for approval, and he gives it, and my wife smiles at me, tosses me a snowball so to speak, even though her hands are raw with the concrete. We are barely able to go to lunch, waiting for the concrete to dry so we can jump up and down on it and express our pride, even though to anyone else our footer can't be anything but a long, dark stone in the ground. But to us it resembles most of what's left of the Roman Empire, and instead of its fall we can think only of its long glory.

15

Joe Coomer died in 1985 and lies now in a small cemetery under a brown stone in Scottsburg, Indiana. My father's father was a carpenter, a coal miner, farmer, factory worker and the only person I've ever loved who's died. His name was Joe Coomer too, and his grandfather before him was Joe Coomer and his grandfather before him was Joe Coomer, none of these men naming their sons after themselves, but the sons in love and pride returning, rearing, the namesake. But we continue to die. Although the name has never ceased to be called, we keep on dying.

I never saw him use the backrest of any chair. Even when he rode in the passenger seat of a car he leaned forward, his face inches from the windshield, forearms on his knees. I don't know what he was expecting. I think he would have accepted anything that came along. It's hard to say because he is of almost mythological proportions to me, as father of my father, almost unbelievable. He waited on us, for one thing. We were the grandchildren of summer. I wouldn't have known him in a coat. I knew him in overalls and white T-shirt, sitting in his lawn chair on the front porch, a Camel between his stained fingers. He watched over South Fifth Street in Austin, Indiana, as if it were the Champs Elysées. I don't imagine any more than three or four cars passed during an hour's time. But once every summer one was our car. We came off the Interstate from Texas or California or New Jersey, drove the mile east into Austin to a three-bedroom frame house with a brown man on the front porch. I don't think he ever recognized us immediately. It always unnerved me, us pulling into his gravel driveway and him still not recognizing us. But when it came to him, who we were, it was an explosion, his hands raised high above his head clapping and his whole body coming up out of the metal chair, leaning first on the arm of the chair, then on the brick porch pillar, and on our car when he got to it.

When I was old enough to know him he was already sustained by his past. He lived in a world as stark with the past as a horse's skull. He was never lost in his memory, but it was as present as his love for us, or the food we ate, or as the present itself.

By the time we were out of the car and he was hugging us, Grandma was on the porch waving us in. Inside, Grandpa sat in his great brown easy chair next to the front door, with my little sister in his lap. The chair was positioned so that he could sit and look out the open door onto South Fifth Street. Five minutes after we arrived, as we spoke, he was stealing glances out that open door, a brilliant, jealous curiosity concerning that street, what might come down it. If a street could bring him something as awesome as his own son I suppose it deserved to be a powerful draw. What could be next? The intriguing thing was, I think, he expected the past to come down it—where he used to be, what he was, his own youth. He expected Joe Coomer to come walking along, tossing arrowheads, telling him something he'd forgotten to remember.

After dinner he took my brother and me into his bedroom so we could hold his oiled guns and sit on his bed. The room never changed. My grandmother rearranged the rest of the house every couple of years, but his room remained the same: chest of drawers in the corner, bed under the window and dresser against the far wall with guns above it.

"Cerilda (pronounced, or at least the way he pronounced it, Shrill-dee), feed these children!" I can't count the number of times in a day my grandfather would yell this at my grandmother. He was vastly concerned that we shouldn't starve to death while under his roof. This command to my grandmother, or to us, "Honey, are you hungry?" or "Go in the kitchen and fix yourself a sandwich," or "Take this money and go to the store for a pop." Bingo. We never turned it down.

Austin, Indiana, is so flat a town that it is often under water. The rain simply shrugs its shoulders and puddles. We had sunny days there, I know, but the air always seemed to be thick with water. The dogs of Austin were always wet; the ditch in front of Grandpa's house always full and green. It was the great fear of our young lives to be pushed into the water that never moved. In truth I was afraid to walk those streets at all. My fear was the result of a combination of Grandpa's vigilance over South Fifth Street, the guns in his room, the stories he and Grandma told of the crazed, doped-up, long-haired teenagers of Austin. The streets were narrow and pocked, and cars came down them at tire-ripping speeds, and there were the ditches on both sides to be thrown into, the waiting, silent water. But we were never bothered by the teenagers of Austin. What remains with me now, more than the occasional brushes with fast cars, is the stillness of the town we walked through. It seemed that everything was somehow lapsing. The yards were quiet; when curtains parted, the faces that appeared were ancient. With each passing summer every building, house and shack seemed to lean over a few more degrees. During one vacation there in the seventies, three houses burned over a two-week span. The only factory in town had closed finally, and the residents couldn't sell their houses, so they burned them for the insurance. I have seen more three-legged dogs in Austin, Indiana, than in the rest of the world combined. I think Grandpa hated this town, its flatness, the houses close around him, the brown, muddy creeks that flowed nearby, although he lived there for almost forty years. He never grew accustomed to it. He knew there were better places to live a life.

My grandfather had followed his two oldest sons out of eastern Kentucky to southern Indiana, from his and his father's farm to the lot on South Fifth Street and a factory job. It was a thing he had to do for his family.

Sustained by his past.

In the late evening, after supper and the ten o'clock news, after my grandparents had retired to their bedrooms, my brother and sister and I would stay up with popcorn and watch TV. Our beds were the couch and floor in the living room, next to Grandpa's bedroom. Our blue-washed, salt-and-buttered summer nights were punctuated with the squeals of Austin's tires from without and the caustic screams of my sleeping grandfather from within. Each scream was a heart attack, a man with a gun on the front porch, a horrible dog-maiming wreck on South Fifth Street. He roared, accused, and we would look up at each other, or rather Phil and Sally would look up at me. One night he screamed, and we settled back into the movie, subsided back into the blue light, and he came suddenly around the corner from the hallway, grabbed me by the ball of my shoulder, and my breath fell out of my mouth like grapefruit. When I could turn to look up at him he shook me and said, "Gotcha." He was in a T-shirt and shorts, and his eyes were like broken eggs. They ran all over his face, and his face wore every expression and none at all.

My mother was at the living-room door then. She held her robe together with one hand, leaned with the other on the door post and said, "Joe, go on to bed now." I started to get up and then I shriveled. She wasn't talking to me.

Grandpa said, "I'm so glad y'all are here, Linda, honey."

And Mom said, "Time for bed, everybody. Goodnight, Joe."

"Goodnight, children," he whispered, and my brother put his popcorn away under the coffee table, saving it, and turned off the television. Sally stood up, my father's T-shirt reaching to her ankles, and went to bed with my mother. And I can't remember now if Phil and I talked, or did not talk, or whether Grandpa screamed again later that night, or slept peacefully.

In the morning Grandpa was not there for breakfast. When he showed up later my father said, "Come on, we're going for a ride," meaning my brother, sister and me and Grandpa.

And I said, "We don't want to go, Dad."

And he said, "Get in the damn car."

So we did, taking the long back seat, holding our hands in our laps. We drove past the houses my grandfather built and helped build during the times he was laid off from the factory, drove out of Austin, out into the flatlands of corn and soybean.

"I can smell a teaspoonful of liquor on your breath, Daddy."

Grandpa didn't turn his head. He leaned forward, gazing out through the windshield.

"Well, I just need to be dead."

"Don't," my father yelled, "don't say that in front of these kids. You don't say that."

"All right."

"You promised me," my father said.

"I know I did, Rufus. I should be dead. I should get me a pistol and shoot myself."

We drove through endless cornfields.

The next day Grandpa was at breakfast eating an egg in four bites. We were going to Kentucky, to his and my father's home place in the Appalachians. Grandpa and Dad were so anxious to be there, they were mean to their wives.

"All right!" my mother said.

"All right!" Cerilda said.

They came off the front porch with purses swinging, resigned to their day-long, back-seat, inter-state fates.

Our car brimmed with the dead on our trip to Kentucky: the old dead, the new dead and the almost dead. The robust held no interest. My father would ask about one of the old Kentuckians, and my grandparents would say he was long dead, recently dead or practically dead, and my father would look up into the rear-view mirror and tell us who this person was to him, what he was to us and, on the average, three anecdotes that may have been intimately or remotely connected with the person. He would do this, and by the third anecdote my mother would have to tell him to take his foot off the accelerator, or to stay out of the median. And so we would slow down and take on the next to die.

Grandpa sat in the front seat, his nose almost touching the windshield, and smiled, and nodded, and smoked, and warned that everything would be, might be, changed. It had been several years since he'd been home, and although he'd talked to people who'd been back he had great fears that he might not be able to find his way around.

The entrance to Bryant's Creek, the valley of the homeplace, was across the breadth of the Middle Fork River by swinging bridge. The steel cables were rusted, the boards cracked and gray, the water below muddy, slow. Gunshots rang from the house across the river. The bridge came alive with us, my brother's bouncing, but we reached solid ground at last and turned in the direction opposite from the house with the gunfire, down a dirt road. We left the women at one of the houses along the river, with a distant relative of my grandmother. I remember my grandmother saying she didn't care anything for seeing that old place, that there wasn't anyone left there to see.

We walked a long way.

We scavenged the first abandoned house we saw. Phil and I picked over the piles of trash, the mounds of bottles, returned with a broken pocket watch and papers from the mid-fifties. We passed other vacant houses, but Dad wouldn't let us go in. In a deep ravine, among rusted barbed wire and automobiles, we spotted a huge steel safe, the door shut. We could see no secure way to bring it out and wondered out loud what treasures we'd decided to give up.

My grandfather stopped, breathing heavily into that knobby country, showing us pawpaw and hickory, vainly seeking ginseng. He looked like a man who wanted someone to tell him what to do next. There were big trees where fields should have been. He leaned on these trees when he could, and once my father told him to sit down on a big rock and he did that. We stopped every few hundred feet so he could rest, and while Phil and I pitched rocks he would curse his one bad leg.

"They've let all this country grow up," and, "Somehow they've got a bulldozer in here and widened this road," and "I knew this leg would do this."

But after a while, the closer we came to the Big Bottom and the hollow of their home, the more he got over his first tiredness. And my father, in his late thirties, ten years away from a double bypass, came into his. My father would ask his father if he remembered certain things, the turn of a road, the look of a leaf in spring, a gas pipe shoved into the bottom of a creek, and Grandpa would nod, mouth open, nodding in the same way he would have if my Dad had been nine. His answer wasn't always yes, but it was always reassuring. I was ten years old before I realized my father was also a son, still a son. For some reason it was a humbling realization.

We broke into the Big Bottom in a dirt-road summer sweat, into a flat of corn and dust with an old church set in the middle. The corn was so tall my grandfather thought at first that the church was gone, but we stepped into a furrow and the spire lay at the end of the row like some lost obelisk.

"I used to walk to this church, and it was a school too, through this field and fill all my pockets with arrowheads. By the time I got there . . . " my father said, and then he bent down to the furrowed dirt and picked up a broken arrowhead. "See!" he said. "See?" and he displayed it in the palm of his hand, showing me, then Phil, then Grandpa, and then he pocketed it. "I used to find so many of them I'd throw them at birds."

"I wouldn't even stoop over to pick one up," Grandpa said.

My little brother looked up at him, terrified. "If you see any, tell me. I'll pick them up, Grandpa." Like dogs, Phil and I went down the furrow nose to ground for Injuns but lost them in the shadow of the old board church. The small frame building was empty, up to the raised area behind the pulpit, which was stacked with bales of hay to the joists.

My father and grandfather stepped through the building, telling us about the hundreds of teams and wagons that used to come here and stand outside on Sundays.

It was still a fairly tight structure. All the doors and windows were out but the roof and floor were still solid. It wouldn't have taken a day to dry it, weather a congregation through a storm. But the sermon was now the collision of dust mote and wasp, the choir in the ears of corn. When we walked out the sun hit our faces as on a Sunday morning.

"The old house is over this way," my father said, and we all followed him down a dirt road and then into the corn again. The field ran up into a narrowing hollow and finally the corn stopped. We banked Bryant's Creek then till it came into another hollow, hip-high in brush and a crop of young trees.

"This couldn't be it, could it, Dad?"

"I don't know, Rufus. I don't know where it is. They've let this good country grow up." He was winded again, and frustrated, and stopped occasionally to curse his leg. "I don't know why they'd let this country grow all up."

"It's too small a piece of land to use a tractor on," my father said, and then, "Yes, this is it. I remember this bend in the creek. This is Doc Hollow. An old doctor lived up here long before we did."

We turned off the bank into the brush and saplings. We crossed a small piece of bottom land and started up the side of a knob and finally found the house there, hanging onto the mountain by allowing trees to grow up through the front porch. All of the windows were broken and there were holes in the shake roof. The front door lay flat on the porch. We walked into that house in August, so it didn't seem as rotten as it did fractured. The exterior walls were also the interior walls, random-width lumber laid horizontally. In places newspaper wallpaper still clung to the boards. The wood had turned a uniform grey except where rusted nails struck down blackly. Two rooms in front and a small ell off the back where the floor swooned from the remembered weight of a cookstove, a spoon in the corner, and that's all there was. I picked up the spoon and asked my grandfather if he thought it was his.

"I don't know, honey. Take it home and ask Cerilda."

The spoon was flattened and green, and although I meant to keep it all my life I don't now know where it is. I can't tell you how this discourages me. My grandmother didn't even recognize it.

My father and grandfather walked through the house silently. I don't know what they expected. Grandpa leaned against the walls, the window sills, the door posts.

In the next moment we were at Hugh Bryant's Cemetery up on the top of a hill. Hugh, my great-great-great-grandfather, was there and my Aunt Bonnie, who died when she was three days old. My grandfather cried for her, and my father for my grandfather and the little girl. Phil and I stood away, wandered through the old blood and stones, and gazed at the fresh grave of a Korean War veteran. The name was familiar. The old papers and watch we'd found in the morning belonged to this man before us. He must have abandoned the house years before we rummaged through it, though.

My father found us and spread his arms wide, "This is all your family, boys. All these people would have loved you if they'd known you."

And my grandfather, "Rufus, we better start home. Cerilda probably wants to go home by now."

Back in Austin, Grandpa took his chair by the door.

"This nasty old town," he said.

"I could have stayed a week," Grandma said. "It wasn't me that wanted to come back."

In later winters they traveled to Florida, and it was always my grandfather who cut the season short by a week or a month to return to Austin. They came to visit us in Texas for a week, and after the first meal Grandpa was saying he needed to be home. My brother and I took them out to the airport at the end of the week. At every crack of the loudspeaker he jumped and tried to board the plane.

"Not yet," we told him, "not yet."

I think he loved his little town.

He knew it like he knew his own bedroom or the view from his big brown chair. He knew it almost instinctively, in the same way that he pulled a cigarette from his overalls, lit it and smoked. In this town he could tilt his head in the way of a bird. He was part of its lore, its build, its power and rumor, and he could affect and change the whole town just by the way he sat on his porch. He could say, "I worked here—built this house—weeded this piece of garden—schooled these children—walked this block to a store—got drunk with this man—hunted with this boy—watched this rain fall." And it all held some marvel for him, as it does for me.

My little brother once pulled a sparkling three-pound bass out of a four-foot-wide brown ditch outside of Austin, and my grandfather was so proud and fond of that fish and its unimaginability that I wished I had been the bass and not myself.

He took me to one of the few remaining pawpaw patches around Austin, and although he was in his late seventies we both picked up fruit as if we were stealing it.

He leans toward me even now with his big, vein-crossed ear and says, "What, honey?" and I say it louder at him, and he responds with the enthusiasm of a rich long life, all over me with love and wonder.

Come back, I yell, come back. Lean toward me and whisper my name. Scare the wits out of me and let my little brother hold your watches and guns. Austin isn't the same, Grandpa.

Look at me, I say, and I hold myself up as a fish, a line through my heart, water in my eye, look at me. I'm a carpenter. I'm building a house. I'm just like you.

16

In late January our house begins to rise above the lay of the land. For a time it is only something to trip over rather than to fall into, but before long the foundation becomes sort of a mini medieval fortress. Wesley Spry builds it. Wesley is a stonemason, and we hire him and his helper to lay the concrete block on top of our footer. (We hypothesize that if we hire professionals at every other step in our construction, they can fix, cover up, anything we have screwed up in between.) He works over a weekend, and I sit on a mound of dirt in the center of our house and watch him work. I sit there under the pretense that I can answer any unforeseen questions he might have about my foundation. The first thing Wesley wants to know is who built our rock wall. "It's been there for over four hundred years," I say. I do not blink. Wesley trades at my father's lumberyard, so all he does is rest his cinder block on his belly for a moment before going back to work. Yeah, I think, anybody can lay a rock wall when all the rocks are the exact same size and shape. I've really only hired Wesley for the half inch of cement between the concrete blocks, the half inch that makes the whole wall plumb and level.

It's amazing how easy a job is when you hire someone else to do it. The foundation rises without any cursing or even sweat; I fold my arms and joke, kick at an anthill, chew a stick of gum. By early Sunday afternoon I have a retainer wall for my foundation. It is six feet high on one end, tapering down to an eight-inch block on the other. We leave the uphill end open so we can drive the truck and tractor in and fill the interior with dirt to the level of the blocks. But before we start, after Wesley and his helper clean up and leave with my sweat in their back pockets, I climb up and walk the parapet of my foundation. This, this, is a fort. I could hold off a horde of Vikings with a slingshot from here. It's almost a shame to fill it all in and cap it with concrete, build a structure on top that any catapulted fireball could knock out in a stroke.

I am sure many government agencies have looked into my past concerning my acquaintance with Mark Laughlin. I have received several phone calls over the last ten years when upon lifting the receiver and saying, "Hello?" I hear a faint "click" at the other end of the line. Mark Laughlin is responsible for these, as well as for the unending stream of motorists along highways who pretend to have flats, and also for the current vogue of wearing sunglasses, implemented by the government so I won't have the foggiest idea who's really watching my every move. I am sure all of my novels have been scrutinized and interpreted to be covert messages to Laughlin and his ragtag group of followers.

But I want to say, admit, whatever, that I personally do not know what the man's politics, faith, religion, sexuality or employment are. It's true that I've known him since Spanish class at Boyle County High School, sophomore year. I remember him appearing in the back row, trying vainly to pronounce, *"Viva la revolución!"* over and over again, till he shamed himself to silence. Mrs. Sharp, our teacher, had red hair, and so Laughlin loved her. It was a vain love because she was married, but it was the kind of love he sought, or so it seemed, through college and on into our roommate years. He fell for redheaded women he could not have. In between these passions he groped through *The New Republic* and his other mail, catalogs for booby traps, army-navy surplus flyers, "Excursions to Nicaragua" pamphlets. He even tried to sell Can-Am motorcycles for a time. Ah well, man is born into a world he did not commit.

Laughlin's style changed drastically a summer or two back, though. A redheaded woman said, "Yes," and Mark was undone. He was married within two months, and although he had sworn off children, she was pregnant with twins within three weeks. His constant search for an alternative lifestyle had finally struck him with a diaper directly between the eyes. But the boy makes me laugh, and so Heather and I invite the family to move back to Texas from Kentucky and help build our house. It is a surprise and not a surprise when they accept. We sit on our couch, Heather and I, after their phone call accepting the invitation and try to repent of what we have done, but it is too late. They are on their way in a Volvo, and our guilt is obvious, unredeemable, unrepentable.

When they do arrive, falling out of the car like wads of tissue, we find that Tina no more understands Laughlin than we do. We had always termed him inscrutable. Tina calls him nuts. She calls him nuts and whacks him on the back of the head and breaks up laughing. Their year-old boys say, "Dada, dumdum." I realize immediately that the world is as it should be. A stability has been reached. Mark even brought his own hammer.

The next morning we begin to fill in the void of the foundation. There is nothing like the uplifting experience of moving dirt from one spot on the planet to another. Now that Mark has arrived, Heather refuses to lift a shovel. She turns from grub to queen with the flick of her wing. As we work she walks the parapet and takes snapshots. She is on vacation.

I have always been a borrower and a lender—especially easy when one is young and has nothing to lend and everything to borrow. My father is my chief source, but I have been known to wheedle from grandparents, uncles and cousins, machine and cabinet shops. So to make our job easier I borrow my father's tractor again and his two-ton lumber truck with dump bed. We decide to kill two holes with one shovel, digging dirt from the pond to make it deeper, and leveling up the foundation with the dirt. We are a precision team. I take the tractor, scoop up six buckets of a fine, black loam and drop it onto the bed of the truck, and Mark drives up the hill, backs into the foundation. I follow him with the tractor so that after he dumps the dirt I can smooth it. We are just as good at this as we were with Tonkas. Construction pride. I want some girls to walk by so I can take off my shirt and sweat in front of them. Mark engages the hydraulic dump, and I get off the tractor and stand by the truck to watch. The bed rises to half tilt, but the dirt doesn't move.

"Go up with it," I yell.

The hydraulic piston extends, the bed rises to almost forty-five degrees, and a single clod drops to the ground. The dirt clings to the steel bed.

"You can't do anything right," I yell at Mark, and I hit the dirt with a shovel. A shovelful rolls off.

"Let me pull forward a bit, back up fast and slam the brakes," Mark says.

"Yes," I say. "Yes."

He pulls the big two-wheeled truck forward, throws her into reverse; the wheels spin back, the truck lurches back and Mark brakes hard. Almost two shovelfuls fall off.

"Yeah!" I scream. "Do that again!"

We do this several times, till the clutch and brakes begin to smoke. Then we stand beside the truck, looking up at the tilted bed, the tenacious earth, then at each other.

"I hope no girls walk by," I say.

Heather taps her foot on the parapet, waiting to take the next snapshot of progress. So we grab our shovels and run at the tilted bed, up the mounds of earth, do a quick spin and thrust and then ride our shovels, pushing dirt, back down to the ground. The bed is at too sharp an angle to be stood on, so we have to work as fast as we can as we fall. A few minutes of this, a couple more clutch and brake jerks, and the bed is clean. Level the ground with the tractor, stand back and despair. Perhaps only another fifty or sixty loads will be necessary to fill the foundation. Back to the pond.

For two days we shuttle back and forth, making a mountain of the ocean, turning back time and erosion. We have become quite proficient at rushing dirt and fighting it down to the ground. Jack rabbits have begun to come out in the early morning and late evening to watch us assault the truck's bed, building the biggest den they've ever seen. Mark and I are just happy that we're not building in a subdivision, where other construction workers would see us. A B-52 Stratofortress from Carswell Air Force Base has been making low passes over us. I don't know if they are interested in Victorian house construction, are keeping tabs on Mark or are using us on a practice run. This is unnerving to a degree. I suppose the pond already looks like a shell crater to them. And two guys sliding down the bed of a lumber truck in the middle of the countryside must seem an enviable target.

During the late afternoon, after twenty or so loads ladled and leveled, I commit the unthinkable. Taking the tractor into the foundation, leveling the dirt, I kiss the concrete block wall with the steel bucket. A block wall is remarkably strong vertically. It will hold up a house. But you can bounce a rubber ball off the side of one and crumble it as if you'd tossed a telephone pole through it. I knock thirty blocks to the ground. I kick the tractor, swear at the fate that put me on the planet and send Heather into the lumberyard for mortar and trowel. "How hard could fixing this be?" I say, and kick at a block. Already my house is in ruins. I hate, I mean I hate, doing anything twice, even when I didn't do it the first time. I take a hammer and chip off the old mortar from the block, and this helps a little. By the time I'm mixing the mortar my adventure tic has clicked and I am almost having fun. I try to remember everything Wesley did with his trowel and level. But when I've gotten the mortar to the correct consistency and scoop some up with my trowel, I don't even get two feet away from the wheelbarrow before the mortar falls off onto my shoe. This is not a joke. I make three trips before I learn the art of holding a trowel. My shoes would take me straight to the bottom. When I finally do get to the block with my mortar, though, I am a whiz. Block and mortar are forgiving. If you don't get them plumb and level the first time, you just take some mortar out, or shove some in. I don't think I could have made a block wall from scratch, but I can patch a hole. It is true that you can tell my section of wall from Wesley's, it lacks neatness and beauty of line, but I think it will hold up a house.

At the end of the second day we wander over the smooth fill, kicking clods, waiting for the funny smell coming out of the clutch to blow off so we can take the big truck back to Dad. The dirt is up to the level of the concrete block now, and it seems strange to walk on our site and walk neither up nor down hill.

Before Mark and I call in the cushion sand to top off the dirt, I ask Heather if a dirt floor wouldn't do. "I can pack it down with the big truck and it will be easy to sweep," I tell her.

The pad sand arrives the next morning, fine and cool. "Just off the beach," Mark screams, and dives in.

A shovel slips into the sand so sweetly it's almost sexual. We pitch it around, sand through the air, and we troll the troop march from *The Wizard of Oz*: "Yo ee o, eeyo ho." Song eases the dull brains of the masses. I had a sand pile when I was a kid, but this seems twice the fun. Perhaps it's because my grandparents, who actually owned the sand pile, wouldn't let us spread it out. It is very hard to spread a sand pile up. I pitch the sand from the pile and feel an odd mixture of guilt and pleasure.

Mark asks the old man driving the dump truck why we didn't get a girl in a bikini with our sand. The old man, who has only three teeth, thinks this is the funniest thing anyone has asked him since he was born. He laughs so long and hard and wide open that I am able to count his teeth many times.

We spread the sand to a fairly even four- or five-inch depth, enough to trench through, lay plumbing in and cover it. The sand won't cut through the copper and plastic pipe and will protect it from the caustic concrete. But before the plumbing and after we finish the beach, we call in the pier driller. He knocks holes in all of our work. He drills twenty-seven twelve-inch-diameter holes in the foundation, ranging in depth from eight feet at the deep-fill end to four feet at the back of the house. The soil comes up sand, then black pond dirt and then our pumpkin clay. Mark peeks over the edge of the eight-foot pier hole and I whisper, "Gophers." Mark is skinny enough to drop all the way to the clay himself. I give him a little nudge in the small of the back, and his limbs spreadeagle in the way of a cat forced into a toilet bowl, even though my nudge barely makes him lean over. I tell him we'll name the house after him. "Every big construction project claims a few lives."

"From now on, I want nets and extra scaffolding support, and a policy that will provide for Tina and the twins," he says.

But by the end of day, after we've pitched over the side the dirt that the drill brought up, Mark decides not to join a union and I promise not to nudge him again.

Plumbing is a peculiar trade. The problems are getting water into a house and then getting it out again. Many advanced societies before our own solved both these problems by not attempting the first of them. A house without plumbing and with a good roof will never shudder in the night, rot, molder or mildew, ruin carpet, walls and ceilings, or even smell bad. A brief, hearty walk to an outhouse allows one time to unzip one's pants or hike up one's skirt. On a trip to the creek or well one might spot a new species of newt. We give these opportunities up for convenience's sake, a shorter walk, no worms between the toes on dark nights. It is a trade, plumbing, based on pressure and slope, on remembering which is the hot side and which the cold. It requires logical thought; quick, efficient hands; and the patience and fortitude of a fly trying to get through a windowpane.

We begin with drainage. (I never dreamed I'd have to write such a line. Drainage. Where are my African lion hunts, my Amazon swims? My big two-hearted river is a four-inch plastic sludge pipe.) And beginning means digging again. We dig trenches of exasperation from one bathroom to the next, to the kitchen, the wash room, the furnace closet, trying to make our ditch slope evenly to one corner of the house from where we'll run a pipe out to a septic tank. We disparage our lives with each shovelful. "Truck it in, shovel it out," Mark screams. "Seems logical to me!" and he laughs, and laughs, till his last laugh, which sounds like the whimper of a frog in a snake's mouth.

The laying of the pipe is less peaceful than this, mainly because we find that two of our ditches don't conform to the fittings made for the pipe. We have to fill them and dig others that divert at more customary, convenient angles, such as forty-five or ninety degrees. Mark and I are appalled that this information isn't printed on our shovels. The world is so complicated. A man can't pick up a spade and dig his own grave without instructions and a disclaimer.

But we finally reach a calm, finish our digging and try to lay our plumbing with some dignity. We work in the tradition of Hezekiah, bringing water to our people. Instead of clay pipe we use PVC, cutting it to size with a hacksaw and gluing it together. We are working with poise, with tradition, slow and with forethought, till I find Mark, ostrich-like, his face jammed into the end of a three-inch pipe. I tap him on the shoulder, because he seems to be having some sort of spasm. He rises, flushfaced, a perfect three-inch bloodless ring impressed around his nose and mouth.

He says, "The glue is wonderful."

"You're not supposed to smell it," I say.

"Aren't supposed to kiss girls, either," he replies.

I take a long, deep whiff, sticking my face into the white, tubular cavern. I come back up, the ring of bliss on my face. "I want to be a plumber till the day I die," I say, which, we decide, wouldn't be very far off if we kept our faces pressed to the pipe of no problems.

The copper supply pipe goes down easily. This is my very first job working with precious metals. Mark holds the end of the roll at the kitchen sink while I walk the rest of the coil down to the well tank, and so on. We'll put the tank inside the house, in the pantry, so we don't have to worry about it freezing during the winter. The pump itself will remain outside, underground. We tie the ends of the pipes that will stick up through the slab inside what will be our walls, about a foot above the concrete foundation level.

Heather comes out in the evening and tapes over the open ends with duct tape so dirtdobbers won't build nests in them. We are very proud of ourselves when we learn neat tricks like this. I don't know why we take such pleasure in thwarting insects, but Heather is almost singing as she tapes over the last open pipe. I think she envisions a bugless house in the sky. She hates ants more than she hates fingers between her toes, and this is a great deal of hate.

In final preparation for the concrete, Mark and I lay steel rebar through the entire foundation in a one-foot-by-one-foot grid, pigtailing it to rebar in the pier holes and to rebar threaded down through the concrete block. This will strengthen the floor by tying one end of the slab to the far end and the slab itself to the deep vertical piers. We then build a short retaining wall (four inches high) all around the top of the concrete block to hold the cement in. This is a tedious process calling for the use of a rented transit to assure a perfectly level slab. With cedar wedges Mark shores up the lumber the necessary eighth of an inch or so, whatever is required, whatever I yell at him with my eye fixed to the glass.

"Down an inch," I yell, "up a half, down an eighth, down a sixteenth." At each of my corrections Mark has to lay down the ten-foot ruler he's holding up, jump down four or five feet to the ground, make the wedge adjustment and finally climb back up onto the foundation to hold the rule stick again. "Down a sixteenth," I say again. Mark lifts the ruler to his shoulder and hurls it at me, javelin style. "You crazy Indian," I scream.

"You're not sincere," he yells back.

"I'm blind," I say. "Ever since I started this house someone has been moving the furniture of my world. I mean, there's no turning back either, is there? I'm trapped in the middle of my life, twenty-eight years invested, hoping for a long-shot payoff. And right now I'm trying to make sure a marble won't roll from one end of a house to the other of its own accord." I am screaming and am probably a sort of purple color at this stage.

Mark scuffs the earth with the toe of his sneaker and says, a bit guiltily, "But it's nice weather today."

It is. We work, finish our day's work, in a blue, bare-branch warmth of a day. There is a breeze from the south, and my shirt is cotton and loose. My shoes, battered from a month and a half of lumber and dirt, are now the most comfortable ones I've ever owned. Heather, lately, runs her hands down my sides and pinches for the love handles, but they're gone. Tomorrow the concrete crew and trucks will finish off our slab, and we can finally get into the life of hammer, nail and saw. I think I am more a wood person than an earth person. I am hoping this, fervently.

After we shim up our last board, we pack the transit away and don't know what to do next but walk around our work several times. The foundation looks very complicated, with its rebar grid, the PVC and copper pipes rising up, the pier holes and deep support trenches, the concrete block wall and its four-inch retainer cap. We step gingerly through it, picking up twigs and tiny clods—not required, but a way to turn it over to someone else. And doing this, and even Mark was doing this, I realize we are showing too much pride in our ordering of the world. We would square and level the earth if it were within our power. What we should do is follow the planet's example and build our houses round, so they'd find their own levels, true mobile homes.

But we become so infatuated with place. I love these oak leaves. I am already nostalgic about our rock wall. I remember Mark, my best friend, heaving a transit stick at me as if it were a Zulu spear. Infatuated with place. The softness of my wife's chest. Things are going too quickly.

"Let's go home," I say, and Mark and I drop into the truck and roll our quiet way to dinner.

The hardest part about building a foundation is the night after you've finished, the night before the concrete trucks arrive: if anything's been left out it will present unimaginable problems later. Heather and I go home to our garage apartment, go to bed, and I put my palms to my temples and press. Your eyes open widely when you do this, if you press hard enough. I stay this way for several hours, trying to remember if I've remembered everything, and finally, when my eyes start to dry out, I give up and fall asleep. In my dreams I chip away at an eight-foot-thick slab of concrete with a rubber mallet.

In the morning it is winter again. A blue norther comes through, dropping the temperature from sixty to forty in ten minutes. We meet the concrete crew at the house site dressed in hats, gloves and down coats.

"Too cold to pour?" I ask.

"Naw," the foreman smiles. "Mind if we build a fire?"

"Naw, go ahead," I say, thinking, if in ten years my foundation and house crack in two and sink like the Titanic I will bury you with your dog in a Coke bottle.

The first truck rumbles in soon, a load mixed with pea gravel so it will fill the holes in the concrete block easily. One of the concrete crew uses a great electric dildo to vibrate the flowing gravel into all the voids. We oversee the crew, four men, in a lordly state: Mark, Heather and I sit in the idling pickup truck, our feet huddled like rabbits under the heater vent. Again I have the guilt feeling for not doing all the work, but it goes away when I think of the $600 check I'll write to these fellows for their day's effort.

The cold thickens, and then winds come; everyone works and walks with their backs to the north. The foreman orders the remaining trucks to add calcium to the concrete so it will dry faster. At the present rate they'd have to rent lights and finish the surface well after midnight. The second truck takes forever to come. The plant is only six miles away.

"I've got to have you come faster," the foreman tells the driver. "Put another truck on it," and he walks off before the driver can answer. The cement contractor has no authority over the concrete plant, other than that he orders from them almost daily when there are three other plants close by. The trucks come in quick succession. The crew backs them up to the foundation, the engine is brought up to high revolutions, and the concrete issues in a smooth grey stream and puddles thickly, spreads thickly, covering all our detailed work. The crew works in knee-high rubber boots, pushing the concrete into place with long tampers on poles. They use a sixteen-foot metal two-by-four to smooth and level the slab. It's heavy work. The concrete moves as easily as a length of chain pushed up hill.

Mark and I get out of the cab after every truckload of concrete and examine the progress. Mark looks at the mixture closely, he's had some experience in testing, and pronounces it acceptable. We press J-bolts into the edge of the slab around the perimeter of the house; they'll bolt the wall plates securely to the foundation. The crew gets back in their truck to warm up, waiting for the next load. And by lunch time the whole house, the back porch and the pier holes for the front porch are done, and we all wait for it to dry. It glistens, and Heather wants to risk her life sprinting across it just one time. I pacify her by letting her press four brass numbers into the wet cement of the back porch: 1987. On the far side of the porch we set in an antique cast-iron boot scraper. It says "Mudders" in a rainbow arc on each side. It will remove the mud of generations. You could grow acres of corn in all the mud it will remove from mine and my children's boots.

Within minutes the crew is on the slab with their rotary smoothers, big paddle wheel operations that knock the high spots down. They work as close to the plumbing as possible and then use a small pointing trowel to smooth around and between the pipes. Then more waiting, then more smoothing, waiting, smoothing. The concrete dries grey, then blues in places. In the evening, after the final smoothing, the sun sets beneath the clouds and sends another glisten across the slab, our floor. We step out on it, strangely, gingerly, as if we're walking on water. For Mark and Heather and me, it's almost magical. Heather takes her sprint from one end to the other, and I jump up and down in place. Even our half-frozen concrete crew smiles. I write them a check, and Heather takes a picture of their eight rubber boots, hung in a row on the headache rack of their truck. Out in the middle of the slab again, I say, "A flat spot. This, this is a place to build a house. Hammer and nail, we start tomorrow."

"Really?" Heather beams.

"Really," I say, and, "Kiss me."

Suddenly, Mark breaks for our pile of tools at the back of the slab, and immediately, instinctively, call of the species, I know what he's after, and I'm with him. We pick up our worn shovels, one apiece, and charge back to the front of the house, the high end of our foundation, and with screams, heave of muscle, we pitch our shovels high and far into the blue cold evening, blades caught in sunset, twirling as they rise and fall.

17

The house on the other end of the long street that reached from Indiana to Texas was at 2213 Refugio, on Fort Worth's north side, and although I only lived there for a few months very early in my life, it is still somehow the draw and conscience and secret of what I'll think next. It's why I still throw rocks at the sun instead of at cats. This was my maternal grandparents' house. They no longer live there, but this is how it was:

My grandmother is at the front door, holding the screen open, and I enter with my brother and sister, and we are no longer human but objects of a complete, all-encompassing devotion. Not only her love, but the house's, the accumulated family of it, a past and present and future sanctioned and protected by a blunt and even arrogant care for one another. "Shhh," she says. "Grandpa's asleep." My mother's father is a fireman and his shift rotates; he'll sleep during the day at times, work at night. We step quietly under the high ceiling into the always cool and dark front room. There are huge triangular pillows there on the couch, and we make forts of them—perfect, air-tight forts that no one knows we're in once we pull the last pillow into place. My mother and grandmother speak softly, muffling their laughter in the kitchen, and we huddle together in the darkness listening to their voices, sucking on rock candy. I remember waking up, great cracks in our fort, Phillip and Sally asleep among the pillows, Mom and Grandma still talking, my cheeks hot till Grandma touches them with her cool hands.

We didn't live in the big house but in the garage apartment on a corner of the largely vacant lot next door. My grandparents built the garage for the car and so that their children would have a place to live during the first few months or years they were married. Mom, Phil and I lived there when the Air Force sent Dad to Newfoundland and the Azores. I remember very little of the actual time we spent there. I remember having the measles there. And I remember stepping out the side door one morning on my way to Grandma's back door and a bee stinging me on my bare belly even before the screen door slammed behind me, as if it had been waiting for me to come out. I remember my belly sticking out much further than my chest. And then the alley and back yard, the things between our door and Grandma's: a sprawling mimosa, a fence easily climbed, the junk man who came down the alley once a week with his donkeys and wagon. More than anything I remember as a very young person knocking every morning on my grandparents' back door, consumed with confidence and in awe of the possibility that it might open and I would be let inside.

They bought their house in 1936, paid $1,700 for it, $250 down and $17 a month. It was built twenty years earlier, during the war, a single-story, clapboard box with a high-hipped roof, along with a streetful similar to it. A six-year-old could bounce a superball inside and not hit the ceilings. I know.

My mother was born on the dining room table (or so they say) in 1940, and soon after Grandma started rebuilding the house. Grandpa did most of the labor, but it was Grandma who got the job done. She ripped out the dining-room wall to make one huge living room for her family, and then put Perma-stone on the outside of the house to deflect arrows. And these changes created the house I knew, along with a couple of swamp coolers hanging from the windows and those cast-iron trailing-vine porch posts that were too easily climbed. But to me, the biggest change she made in the house was installing a wall of bookcases in a hallway between two bedrooms. By the time I was born in 1958 they were full of books. Hooray for her. Hooray for her. I spent half of my time in her house on the floor in front of those bookcases and the other half in the kitchen, where she was. In the summer I went to thirds: one third with the books, one third with her and one third in front of the swamp cooler.

The last time I cried for my parents I was somewhere between five and eight years old, standing at the front window of Grandma's house watching them drive off. It was very early in the morning, and they were on their way to Puerto Rico for a week's vacation. Grandma took me by the shoulder and led me back to bed. When I woke up again, an hour or two later, it was as if they'd been gone for years, and there was nothing left to do but get up and enjoy it. I had awakened in the world of 2213 Refugio, which to me was little short of Oz, with all its sweetness and danger. Grandma initiated us, so to speak, by taking us to Clark's department store (now Billy Bob's, world's largest honky-tonk) for a haircut.

"Tell the barber what kind of haircut you want," she said.

"I want a regular," I said.

"I want a baldy," Phil said.

And so it was. A regular meant three inches of hair on top and a quarter-inch on the back and sides. A baldy meant no hair at all, and Phil was very proud of it. By the time Mom and Dad returned it was like a five-o'clock shadow, with all of Phil's childhood scars outlined in a neat way, and our parents never went to Puerto Rico again. Even I walked around that first week in a kind of wonderful chagrined awe of Phil and his bony noggin. It was a fine target for an older brother to aim at but gave me the willies whenever I took hold.

Oz: a front yard of soft St. Augustine grass, hedges with hiding places along one side of the house, a perfect half-round rock sticking up through the soil across the street. The rock was smooth and grey and three feet in diameter, and no matter how hard we dug around it with spoons, we couldn't pry it loose to see if it was perfectly round. We were sure it was a fallen planet, and so played king of the planet on it for hours. Then we'd race back across the street, hide-and-seek in the hedges, drink a pilfered Coke there, and then tackle on the St. Augustine.

When Crazy Mike stepped out of his front door two houses down, we gave up the street to him. His clothes reminded me of the blankets Grandpa kept in the back of his station wagon to wrap his tools in; torn, brown, oil-stained, draped over him. He wore suit clothes all through the Texas summer. He was a big man, easily over six feet, long arms and legs, but he never took big strides. He shuffled up and down Refugio picking up cigarette butts, bottle caps and rocks, pitching them to the curb. He spoke while he did this; from behind the living room window we could see his lips moving. And once when he came up on us unaware, he yelled something unintelligible as we played on planet rock, then he mumbled off. I always thought he was an old man, he was so grey, burnt and gaunt, but Grandma told us he was younger than she was. She didn't know why he cleaned up the street, or why he talked to himself or why he never wore socks. The only thing she was sure of was that he meant us no harm. We thought this was odd, because it was the only thing of which we weren't sure. We never thought the worst of him; that was too far from our conception of the house on Refugio, but we never trusted him enough to be within reach.

In between all this there was the Oz of Grandma's bookshelves. Encyclopedias, *National Geographics*, *Reader's Digests*, westerns and coffee-table books on the wonders and secrets of the past. I owe a great deal of my books to her. I think I ruined my eyes in that dark hallway, yanking down an encyclopedia and just reading whatever it fell open to.

I'd read, eat, sleep, go outside. She'd open the door for me, saying, "Be good." Grandpa would be outside in the front yard watering the St. Augustine. Mom and Dad finally came home, home from Puerto Rico with shells and smooth pieces of glass from the ocean floor, and our parents seemed to trade us these for our grandparents and their home. We moved from Texas to California shortly thereafter. Grandma and Grandpa were about to sell the old house anyway. They'd slowly been moving their lives to a lake house they'd built over a span of years.

We go into Fort Worth once in a while now to drive up Refugio. Our family lived there for forty years, and it's been hard to let go. The people who bought the house painted the trim orange and let all the grass die, and there are junk cars in the vacant lot next door. Planet rock is still there. Other children play on it. Crazy Mike has died, and there are stones all over the pavement now.

I drive down the street to the big house and suddenly see my little brother up on the porch roof, his feet dangling. He's only four and not supposed to be there, and my heart jumps and stops and jumps again, and then I slow down and remember that it's just my past still living there, vibrant and new, and my love for him and it.

18

Thoreau built his house to a great degree from on-site and local materials. His timbers, studs and rafters were cut and hewn from "tall arrowy white pines" near his site, and the stones he used to build his chimney were brought, an armful at a time, from the shores of the pond below. His boards were bought of an Irishman at a good price, good even considering the fact that he had to tear down the Irishman's house to get them. Although I don't think this fit in precisely with Thoreau's vision of economy and simplicity, tearing down one house to build another. He helped forge some of the house's hardware, and I'm sure the bricks of the hearth and the glass of the two windows were New-England-made.

We chose our site—and fit our house plan on it—so that we wouldn't have to cut a tree down. Our oaks are anything but tall and arrowy. Building stone, when found locally, goes at a premium. And finally, it's more economical to buy new boards than to hire labor to retrieve used ones. Our first bundle of lumber off the truck is spruce, from Canada, and the first nails we use are twelve-penny VC (vinyl-coated) out of a box marked "Taiwan."

The world doesn't work as simply as it used to, though from Henry David's point of view, it never did. It seems the closest source of any of our materials will be the yellow pine forests of east Texas, great-great-grandseedlings of the trees that provided the lumber for the old Veal Station school. We'll buy redwood, for exterior trim, from California, oak for interior trim from Kentucky, and marble for our front entry from Italy; our oak flooring will hail from Taiwan also and our verandah tile from Mexico; and while most of our nails are U.S.- and Taiwan-made, some will arrive from China. In fact, probably every house built in America in the last few years has Chinese nails in it, and if the communists are half as devious as some would have us believe, the dwellings of America will all collapse at one precisely preset moment.

The slab is poured on Thursday, April 2, and early Friday we're out on the big, flat rock in order to catch it green, still soft enough to accept a concrete nail without shattering. I have my own crew: Mark; my sister's husband, Bobby; and Homer Day. Homer has worked in construction for a few years, worked for my father in the lumberyard for a time and, over the past six months, built and sold a custom home. Experience is swell, but Homer's best attribute is that he's six-foot-three and about 215 pounds. He'll be handy when it comes to lifting the two-by-twelve-by-twenty-six-foot joists up to the second floor. Bobby is fairly stout too, but he'll have to go back to work Monday. He's here to help us get started. Mark and I are both rather pitiful physically. We don't weigh three hundred pounds together, and weigh what we weigh in a stringy and obtuse way. Mark thinks our builds ensure a plucky endurance, and I think we will need it, since we have to make two trips to every one of Homer's: the lumber truck, arriving five minutes before we do, has dumped the load ninety feet from the slab. Bobby, who credits the human mind with less integrity than I do, tells the driver to his face that he is stupid. After the driver has left and we've carried lumber for an hour, back and forth, back and forth, I also tell the driver he is stupid.

When we've lugged over enough lumber to start, Homer and Mark begin making corners and tees, the connecting members of the walls. Bobby and I lay out the walls on the foundation with a blue chalk line. And then we all four begin the construction of our big Victorian bird cage. The plate and stud walls go up quickly, with a pristine newness and beauty of line and angle, and Heather and I are almost aghast at the sight of a house being born. The spruce is white to manila with an occasional dark-brown, round knot of color and hardness. It wrinkles the air with its fresh, sappy smell, and I am proud and fond of every board out of the bundle. I am surrounded by the sharp ring of hammer to nail till the nail is struck home and hammer thumps the lumber one more bump. Sawdust, eye-stinging, raunchy with fragrance, at our feet, blowing through the house and off the slab to the ground where it soaks up, darkens. Whine of saw, curse of a nail bent, drop of board to concrete, and Heather in the field as fresh and beautiful as the lumber, taking pictures. A house is going up. The world is on fire, a species will survive, love will abide, the corners are square. I walk through, a hammer in my hand, jangling nails in my new canvas apron like riches. I am a carpenter. I work with wood.

During the afternoon we make our first mistake, a hundred-dollar mistake. Bobby and I get crossed up over the size of most of the windows in the house, and he and Homer build seventeen window headers (doubled two-by-twelves that support the weight of the house over the window) three inches too short. There's nothing like doing something twice to nasty up your afternoon. Fool, I think, what made you think you could build a house? This is approximately the five-hundredth time I've thought this so it doesn't bother me as much as it used to. The sky remains blue.

By evening we've finished almost all the first-floor walls. We'll begin again at seven in the morning. Heather and I walk through the cool evening, stepping through the walls and the outlines of windows and doors.

"The rooms seem small," she says.

"It's because you can see through the walls," I theorize. "You're comparing the whole outside world to our dining room."

"I like it like this," she says. "I wish we could leave it like this and not get wet."

We step to each window opening and look out. "This will be the view," I say, and she smiles. We walk down the hallways, spin in the tower room, and she pretends standing at the kitchen sink. The sink will be in the corner of the kitchen, and we've put countertopto-ceiling windows in both walls. The view is to the south and west, for sunsets, and one obliges us, framed in spruce.

"Maybe we should spend the night," Heather suggests.

"Not yet," I say. "Don't worry. We'll be the first to move in." And we both sigh the sigh of being between. Which is a better sigh than being outside.

Things go so quickly over the next few weeks that Heather can hardly keep up with her camera: hundreds of trips back and forth the eight miles to Dad's lumberyard and hardware store for nails, lumber, flashing, tools, everything we forgot from our first trip there in the morning. I will go, then Heather will go, then Mark and then I'll go again. Every trip is run at top speed, saving daylight, one more board up before some hypothetical windstorm/earthquake/tornado hits. It is the fear of our lives that our stick-and-nail house will blow over before it's structurally sound. We're working on the vulnerable third story of the tower now, some forty feet from the ground. Beneath us, the last couple of weeks: two-by-eight joists which support the third floor, a ten-foot second story, another plywood floor, two-by-twelve joists and then the ten-foot first story. The studs on the first floor are a close twelve inches on center, to help support the weight of the house. The second story studs are sixteen inches on center. The third floor seems solid, but you can feel gusts from your clinched toes to your heart.

So while Homer and Mark and I work on the upper floors, a second crew starts to sheathe the whole house in waferboard—a substitute for plywood (which was itself a substitute for planks) that's cheaper and stronger as well as resistant to water damage. It's a two-man crew, and the men seem to work well together although they've never met before. Don has some experience, but it's Michigan experience; he's sure my house will be an icebox. He's only been in Texas for a few months, looking for work, and asks that I pay him in cash. He lives in the same rental park that Robert lives in. Robert has no experience but is eager and goes out and buys his own hammer. He arrives at the house the first day in a Trans Am, with two women. They drop him off and pick him up at the end of the day. Don is my age but seems ten years older; Robert is four years older than me but acts ten years younger; for some reason they are both interested in how old I am and how I got so rich. I tell them this is a wholesale house, and that I married money. They ask because it is a big, imposing house. The first floor is almost 1,650 square feet, the second, minus verandahs, is around 1,500 square feet, and the third floor is another 1,200. They ask how many kids I've got, and I have to tell them I'm building for the future. I tell them my wife designed it. I have the peculiar sensation, during their questioning, that they're looking up my skirt.

The decking for the third floor presents a problem. The problem is we don't want to carry the thirty-five sheets of three-quarter-inch plywood up to it, especially since we don't have any stairs yet. We've been working too fast to stop to build them; there's also the fact that none of us have built any before. So we make a quick run into town and pick up a spool of nylon rope and buy two old tackle blocks at one of our antique malls. Back at the house our block and tackle work well, after we remember how they go together, but we drop the first sheet three times before we get it up. We try rigging up an old piece of rebar as a platform for the plywood, but it sways and bends; we try tying the wood in a sling, and this works, but it's miserable labor to extract the lumber once we get it up to the third floor. Finally, Dad comes down from his house and ties a C-clamp onto the rope, clamps one corner of the plywood, and we raise it up, unclamp it and drop the tackle back down. The whole floor is up within thirty minutes. I'm not sure if every father is like him—Mark says his is—I mean, having an uncanny knack of seeing every blunder his offspring has made over the last two years as soon as he walks in the door. What's even more humiliating is, he fixes and solves my blunders just as quickly. And his answers aren't ones of experience; my Dad's only had the lumberyard for three or four years. They're answers of insight and ingenuity. Mark and I stand back, lips up alongside our noses, our shoulders slumped.

"We were just about to do that," Mark offers.

"Yes," I add, "if we'd just had another twenty years or so."

Sunday comes and Mark stays home to play with his wife and twins instead of me. I buy a stair square, which comes with a little blue book on everything you need to know about stairs. I sit myself down in the middle of the house to figure it out. It's rough work, trying to think. But this day is almost like sleepwalking. I am relaxed to the point of shuffling through the sawdust. Instead of thinking what's next for five people, I only have to decide for myself. The day is a blown leaf. I wear jeans, a T-shirt and a wasted long-sleeved flannel shirt for warmth.

All my good tools are at hand. I don't know why I have a fondness for them, but they draw me. I doubt I'll ever take them to bed with me, but I do like to touch them, have them nearby. I can play with a level for hours, and a self-retracting tape measure has to be one of the greatest toys ever invented. I own three different types of claw hammers, of various weights and lengths, and the ugliest phrase any of us on the job knows is, "Loan me your hammer." You use a hammer for a couple of weeks and any other feels awkward and clumsy in your hand. I'd just as soon lend my hand to a guy so he could slap somebody with it.

I'd much rather work with wood than with earth. I've never understood these orgasmic gardeners who thrust their hands into loamy ground and bring earth back up, dung- and slug-ridden, so they can give it a good, long snort. But wood, grain of pine, fragrance of cedar, pure stink of freshly sawn oak: the best earth and sky and water can make. My favorite area to work at my father's business isn't in the hardware, although nuts and bolts have their powers of persuasion, but in the lumber bins, stacking and sorting the white and yellow pine, spruce and cedar, redwood and oak. It's one of the best places in the world to be during a rainstorm, when some of the lumber catches the water and colors dramatically, the variety and brilliance of grain, and all the odor inside comes out. I always wanted to take a girl there during a storm, up under the tin roof.

My stairs go steeply but well. I use long yellow-pine two-by-twelves for stringers, notching the rise and run out of them with a circular saw. And then use the too-short window headers for treads, just tacking them into place because these steps are only temporary. The little blue book says if there's a difference of as little as one eighth of an inch in a step's height, you'll trip over the step every time. I suppose this is why we misjudge people so often, none of them being exactly alike. We trip over their morals or hairstyle and then look back at them as if it's their fault. Something inside makes me want to rebuild the stairs so that each step differs: narrow tread, wide tread, short riser, tall one, so that my children will grow up expecting the unfamiliar, and will meet the world unprejudiced. At any rate, a set of stairs of this sort would at least confuse burglars in the night.

When I've finished, floor to floor, I sprint up and down the staircase a couple of times to see if it will hold. Boy, this will take its toll when I'm old. When we thought of the space and coolness of ten-foot ceilings, we didn't consider the four or five extra steps from kitchen to bedroom. But the men will be glad to see it Monday. I can't wait to show Heather, but before I do, I crouch down and step under the staircase and scrunch there in the semi-dark just once, before I give it up to the children. It will be years before they're old enough and brave enough to find it, but I may as well break it in for them.

Grandma and Grandpa and their only son, my Uncle Tom, come out while I'm cleaning up the mounds of sawdust. They check on my progress and yell encouragement almost every Sunday. They are my pastor, choir and congregation; I'm currently engaged in constructing the church. Grandma likes the stairs, but her second comment is upon the need for an elevator. I resolve to accommodate her.

I show Grandpa my fire boxes, built into the first- and second-floor hallways. Not only are Heather and I paranoid about high winds, but also about fire. The fire boxes are just big enough to hold fifty feet of water hose and will be more of a comfort than an extinguisher would.

"Well, I never would have thought of that," he marvels, with his usual self-deprecating praise. I accept it and don't remind him of his forty years as a fireman. Grandma walks slowly up my stairs, holding the bare studs of the wall. She doesn't like being able to see through the backs of the steps. Tommy follows his parents and whacks his head on the two-by-twelve joist that defines the stairwell. He is almost six-foot three. I am five-foot eight.

"Design flaw," I mumble.

"You're not a very brainy kid," he says.

"If you had any hair," I say, "it wouldn't have hurt so much."

"House full of short people," he says.

They walk through our work, looking out windows-to-be, stepping out on the two verandahs. Grandma and Grandpa fold their arms and praise and talk about the future, telling me to be careful and not fall down stairwells or off ladders. Why didn't Thoreau write about his mom visiting him at his hut? You know she must have dropped in. Uncle Tommy's favorite room is the tower, an octagonal structure off the living room and master bedroom. Five sides of the tower have windows, and wind blows through as if the tower were a tree in winter.

Grandma can't wait for the library to be finished. It's a two-story room with a big bay window at the end of each story. The upper story is nothing more than a three-foot gang walk around the walls. We plan to put a railing around the big hole in the floor and bookshelves on the three windowless walls. And a rolling library ladder up there too, of course, so I can climb up on it, grab a brass handle and wing myself around the room to *Heart of Darkness*. I was planning to install a spiral stair from the first floor to second, but I think I'll look into an elevator instead, something simple, maybe a truck winch hooked to an oak office chair. I'm serious.

On Sundays, Heather's mom and stepfather, Eleanor and Leon, come out from Dallas with their camera. They've just gone through a two-year reconstruction and add-on to their home and hope our experience is better. They marvel at how fast the framing is going, and Leon, at age forty already thinking about retirement, comments on how spacious the house is. Heather says, forget it, I'm trucking you both off to a Florida retirement community. As we show them through the house, the week's progress, I wonder if they can see that I don't have the faintest idea what I'm doing. That I'm bluffing my way from day to day. I think about telling them the truth, then consider the possibility that they might take my wife back. But then I realize that they're twice as supportive as Heather is, that perhaps they were glad to get rid of her.

Monday morning we begin the roof, tying the first rafter into the third story of the tower that's already been built, then moving down the length of the house, butting rafters into the ridge board at the peak, and on the bottom nailing them to the two-by-eight floor joists on the edge of the house. We use a twelve/twelve pitch, which is a forty-five-degree slope. The peak is some fifteen feet from the third floor, so to nail the rafters there I stand on top of a ten-foot stepladder and hammer at face level. I cannot remember a windier spring. As I wait for the next rafter, I look out and for the first time see the skyscrapers of Fort Worth, twenty-eight miles away. I can see only the top halves of four or five buildings over the distant treetops of a butte. Suddenly I wonder if the people in them can see me, and I wave. This isn't ignorant. Waving at somebody is generally the only way to find out if they'll wave back. Mark and Homer look up at me, then shake my ladder to get my attention. They're ready with the next rafter. The house is beginning to look like a great tent. The rafters run from one end of the house to the other and are only broken up by two small dormers on each side. Later the roofs over the library and dining room will join this roof, but since this one is so much taller its line won't be broken. We won't build hips on the ends of the house but gables, so we can fit fretwork into them.

While I'm nailing up one of the last big rafters, trying to toenail into a difficult spot, my hammer caroms off the nailhead and lands on the corner of my thumb. I realize this instantly because my thumb, with hand attached, jerks back and slaps me in the face, knocking my spectacles off and almost knocking me off the ladder. There is only one greater pain for a man than hitting his own thumb with a hammer held by his opposite hand. This greater pain is hitting the thumb a second time a few seconds after the first strike. The senses are acutely aware. If someone had told me I was going to hit my thumb with a hammer while building my house I wouldn't have begun. But here I am, only a few months into my house, and I've hit my thumb not once but several times. If you're reading this book with some intention of building your own house, be warned: you are going to inflict an arbitrary, intense pain upon yourself many times.

There are several classes of thumb ca-whomping: the nick with a resulting blood blister; the glancing blow that simply scrapes away the skin; the ricochet (bending over to hit a low nail and bouncing the hammer head off your shin to your thumb); and then the terrible, completely-miss-the-nail-swinging-with-all-your-strength square smash to the thumb that makes you drop the hammer, rear up as a rutting stallion, prance in circles and hold your throbbing member as gently as if it were your genitals. I know why the little plastic shingles on the ends of our fingers are also called nails. I have put half my house together with my thumb as a fastener.

The only class of thumb ca-whomping that doesn't physically hurt the hammer holder is the ca-whomping of a co-worker's thumb. While working on the roof above the dining room and nursery I accomplish this feat, earn this rarest of carpentry merit badges. I'm nailing up a small jack rafter in a corner so tight my hammer bounces back and forth between two boards like a tennis ball caught under the bleachers. Mark and Homer are both looking on, helping hold the board. I pull up for a split second, trying to get a better angle on the nail, when, inexplicably, Homer reaches in and takes hold of the nail in the middle of my swing. I smash his thumb like a bug.

"That didn't hurt," he denies.

Mark and I look on, aghast. We protect our own thumbs, vicariously, instinctively, by putting them down between our thighs, with the other important things we own.

"Why'd you do that?" I ask. "God, I'm sorry," I say.

"I thought you were giving up," Homer says.

He has stepped out of the low rafters onto the open area of the third floor. He keeps holding his thumb up to the light and staring at it, as if it were an old coin. Then he says, "Goddamn, that hurts," and he drops his nail apron. "I've got to get a drink of water." Mark and I follow Homer down to the first floor where we find him sprawling on the concrete, panting. "It made me queasy," he says.

"God, I'm sorry, Homer," I say again.

"It's okay. Just lay your thumb flat out on the floor here and give me my hammer."

My whole body woggles as if I've just found an earthworm under my tongue.

"Anything else," I plead, and run away.

After we finish with the rafters and stud up the gables we help Don and Robert with the sheathing. The house gets darker and darker inside as the waferboard goes up. We'll no longer walk through it and see the world fractured into geometric shards. We'll see it as rectangular landscapes, an occasional portrait, framed by our windows.

At end of day I pay Don his usual draw. He and his family seem to be making it from day to day. He's been looking for full-time work but finds only jobs like the one I've offered, jobs that last a month or two. Over the past two weeks he's missed four days. His truck has broken down twice, and twice he's gone into Fort Worth to see if his unemployment check has come in from Michigan. He's gotten to know Robert better than Homer, Mark or I have, and tells us that the two women who drop Robert off and pick him up are both his. One is his wife and one is his girlfriend, and they all three live in the same house.

"No shit?" Mark says.

"No shit?" I say.

Even though we're both married now, Mark and I both come from lonely-guy backgrounds. We were lucky to have two dates in one month, much less two women in one house. Our respect for Robert grows. These women certainly aren't after him for his money. And we consider his looks and decide they're not after him for those either. Then we cannot figure anything out and conclude he is from another planet.

Robert waits till the end of the week for his check but asks for Thursday off. I say, "Sure," right in the middle of his explanation, then I realize he's telling me he needs off so he can go into town and check in with his parole officer. "Sure," I say again. "Do anything you like." None of the four of us has the nerve to ask him what he did, but we surmise it was drugs, since the police are always raiding his rental community for that.

"You've hired a convict," Mark says to me on the way home.

"I'm an equal opportunity employer," I say.

"At least it explains the way he swings a hammer," he says, and it takes everything I have in me not to rise to his enigmatic bait.

In the late evening Mark's wife, Tina, calls. She tells Heather that Mark has just found four eight-penny nails in his underwear. We'd worked all afternoon without our shirts on, our nail aprons hanging from our belt loops. Mark must have dropped the nails down into his pants instead of back into the nail apron. It was now eleven o'clock at night. He'd worked, driven home, eaten dinner and watched TV for some three hours with the four two-and-a-half-inch nails snuggled in his shorts. Heather and I giggle ourselves to sleep.

In the morning, Mark doesn't say a word. This is usual, but I expect some sort of remark, excuse, at least. I finally confront him and he asks, "Did Tina call you?" I say yes, and, after torturing his lips for the briefest moment, he assumes an immodest pride.

"Balls of steel," he says.

"No," I scream, refusing him. "They were body temperature. You never had the least notion you were carrying them. You're a leper. Your brain never leaves the confines of your skull." And I get out of the truck, slam the door.

"Balls of steel," he says. "I had a project at home."

"Yes," I whisper, exasperated, beaten down. "It's lucky you didn't get an erection yesterday. You might have nailed yourself to my house."

Robert and Don come around the corner of the house then, pretending they haven't heard what we've just said.

"Great," I tell Mark.

"Do you think he was in for rape?" he asks.

"I think he likes the way you walk," I say.

I put Don and Robert on the grunt detail, picking up all the scrap and trash in and around the house and then carrying waferboard up to the third floor so we can start sheathing the roof later in the week. I hate to give up the cleaning to them; it's one of my favorite things, transforming a pile of sticks into a swept, clean space. It makes me think there's progress occurring.

Homer has left us for a better job. I don't think the thumb-smashing had anything to do with it. I hate to see him go; we lose his experience, his strength and his generally carefree attitude. At one point during the construction of the front verandah we ran up against a technical problem, and as my brother and I discussed it Homer took out a cigarette and smoked. Finally we asked him, "Well, what do you think, Homer?"

He turned back at us from gazing out across the green of the farm and replied, "Well, I like to take each day as it comes. I don't know. You just try to get along in life as best you can. Don't hurt nobody, feed your kids. But I'm pretty happy with it." Phil and I just looked at each other and smiled.

Then Phil said, "Me, I like sex as often as I can get it."

Mark and I take ourselves and sixteen eighteen-foot two-by-six spruce rafters up to the third floor, along with all our tools, two sawhorses and a ten-foot ladder. We're finally going to build/attempt to build the peak on the tower. Heather has come out with her camera to record the event. We'll be working on a platform on top of the third story of the tower, some thirty-eight feet off the ground. We'll have to stand the ten-foot step ladder on top of this to reach the peak of the tower, some sixteen feet above the platform. The wind is blowing but is a little less gusty than usual.

I understand the very rich build their towers on the ground and hire helicopters or cranes to lift them into position. This is the coward's way out. If I'd wanted a prefabricated, bolt-together house I'd have ordered one from Sears. I do begin to wish I were rich, though, because once we get all the materials to the third floor, we stall. We don't know what to do next. It's sort of a daunting project, and although Mark hasn't admitted it, I think he has a healthy respect for heights. For an hour or two we try to get around sawing and hammering by drawing triangles and octagons on scraps of lumber, trying to remember our geometry and trigonometry from high school. Mark remembers, or reinvents, better than I do. He was always better at that sort of thing in school, walking in on a chemistry test without having studied and rediscovering the formulas, solving problems through back doors. We finally succeed by throwing away the blocks of wood, clearing a large area of the plywood third floor and drawing a life-size section of the tower roof on it. Everything about our house goes slowly because everything is a first time. Mark and I are sure we have built two houses already, one unacceptable one and one mediocre, passable one to which we continue to add. We cut the first two rafters and take them out to the platform through the four-foot door that connects the main third floor to the tower third floor. (We could have put a bigger door there but it would have messed up the line of the roof, and we were building the third story of the tower for adventure anyway.) Then we angle them up on the platform and pull the ladder up. Seeing no way to nail up one rafter at a time, we put them together on the platform with a strengthening crossmember, and I begin up the ladder with it, an eighteen-foot letter A. Mark holds the ladder while I inch up. Spruce is fairly light lumber, and two long boards nailed together at this angle have a life of their own in this wind. I reach the top of the ladder and feel like a balancing toy. Finally the birdmouths (notches cut out of the rafters to enable them to sit flat on the wall plates) pop onto the plates, and the weight is off of me. The aluminum ladder sways, the rafters sway, and I sway with them, while Mark frantically nails the rafters to the plates. Then we look at each other.

"What now?" he asks.

"I can't let go."

"I better go cut another rafter," he says.

I don't have to say, "Hurry!" but I do anyway. I say it twice, "Hurry!"

I'm standing on the very top step of the ladder, the one labeled "NOT A STEP." I'm not afraid of heights, generally. I can look down into the Grand Canyon or off the Eiffel Tower without any trouble. These places have iron railings that don't move. My problem here is that the railing moves, and my four aluminum feet standing on a piece of half-inch plywood below would just as soon move with it. The skies are cloudy above, and I would be so brave as to shake my fist at them if I were brave enough to let go. I'm up above the treetops by a good twelve feet and can see all of Fort Worth clearly. While I'm shaking, Mark finally arrives with another rafter. We raise it into place, nail it top and bottom, and I let go, my hands still lingering near the peak as if I've just let a bird I've built fly for the first time. We add a few more rafters, and Heather takes her first picture. She is far away, down on the ground, and Puppy and Butch play at her feet.

"It's a rocket ship," she screams. "Or a Christmas tree."

"It's a teepee," I yell back at her. "Tonkawa homage."

The dogs look up at me, square off their limp ears and bank their heads.

"Woof," I bark at them, and I raise my hand high and add, "Master!" They leave Heather, and I hear them entering the house and bounding up the flights of stairs. It's nice to see them. I mean, it's nice to see them having so much energy and room to run; their yard is small back at the garage apartment, and I'll be as glad for them as I'll be for Heather and me when we move out here. But this coming upstairs for a quick lick and hello has to stop. Every time after they say hello, they walk back to the stairs and begin to cry. These are grown dogs, seventy and ninety pounds, and several times a week, sometimes several times a day on their prolonged visits, Mark or I or one of the others has to carry them back down the two flights of stairs. I have never been more ashamed of a dog. I pick Butch up and call him Fifi all the way downstairs. A big dog doesn't like to be carried anyway, and so I try to add to their chagrin as much as possible, hoping they'll attempt the stairs themselves. Then back up the stairs for Puppy, still whining, and on the way down I tell her I saw her shitting in the woods earlier. Then I feel bad, because Puppy has such sad, big brown eyes, and so I have to play ten minutes of chase the dried up paint roller with them. I don't know what kind of father I'll make.

Just before we quit for the day, Don's wife pulls in and tells him his best friend has just killed himself. "He hung himself," she says, within earshot. "Mike found him, but he was already dead."

Don says, "Oh."

That's all I can hear. A few minutes later he comes over and tells me this man and his wife had just split up, and that he would need the next few days off. "He was my best friend," he says. "They came down from Michigan with us. He's tried before but never done it well enough."

I look at his face and have nothing to say, nothing to do, but wonder why I am not him, why I'm not this other human standing in front of me.

I come home at the ends of days, wash my hands, kiss my wife's face. We make dinner and eat it in front of the TV, fending off Blossom and Eliot. I tell her what a fine job we've just done, or that we'll never finish. The cats play with our shoestrings, and we attack them as giant spiders, watch them freak out and explode across the apartment, bouncing off walls, cabinets and each other. We surround ourselves with our best friends. The dogs come in, and our rabbit, our first pet, Bumby. The cats take to high ground, more afraid of the rabbit than of the dogs. For he is a strange one to them, this lop-eared creature. The dogs will lie down with a cow-hide bone, but this rabbit lives for electrical cords. He searches through our apartment till he comes upon one and nibbles it into one-inch segments whether electricity is running through it or not. We think he's become immune to the power of the current, through a fantastic and often repeated series of self-inflicted shocks. We use natural-gas room heaters in our small apartment, and they dry the carpet to a buzz. Bumby circles on the floor, building up static electricity in his furry rabbit's feet till he is shocked and immediately is launched three feet directly up into the air. The cats act as if they've been shot with a water pistol when they see this, and even the dogs are impressed. Humans screech with sadistic amusement.

Bumby has grown since we got him, and when we move out to the big house we plan to build him a bigger cage.

Over the next week, waiting for the spring winds to die down so we can roof the house, Mark and Robert and I build a front porch. We bolt a two-by-eight to the concrete block of the foundation, set posts and beams across the concrete porch piers and lay two-by-six flooring. The porch is eight feet wide. It spans the whole front of the house, wraps around the tower and runs down half of the south side of the building. It's a beautiful thing, new spruce at angles, and looks like a deck, since it doesn't have a roof above it yet. We'll wrap lattice underneath to hide the pier and beam supports later. There's a good deal of room below it, almost six feet of height in front, and before long I'm going to throw a spoon, a broken toy, an ice cream stick and four steak bones under our front porch, just to break it in, make it look at home, well-loved by child and dog. By building a front porch I also construct a fine doghouse, refuge and shade. I think there will be some days when I'm down there with Butch and Puppy, cool, away from the world and the feet of first and second floor, whiling away the hours with a book—my dog bone—gnarling covers, dog-earing pages.

19

I remember, more than anything, the heat of the pavement upon my feet. My memories of our house in Saginaw are located in an endless Texas summer of blacktop-skinned baseballs, banana-seat bicycles and cicadas among cottonwoods. We lived there for six years, from June of 1964 through November of 1970, in this suburb of Fort Worth due north of the city. I'm sure it must have turned cold at some point, I must have worn shoes while ice draped the trees, but my feet don't remember it. All I remember is running to reach grass, shade or a bicycle before my bare feet stuck to the frying pan of the roads, driveways and sidewalks. "Boy, it's hot," I would say.

We moved from Grandma's garage apartment to 536 Opal Court when my father left the Air Force and took a job with IBM. One of the first things my father, brother and I did was to make and hang a banner welcoming my mom and new sister, Sally, home from the hospital. She came home to an enlarged A-frame house: a steep roof that encased the second story and sat on the walls of the first, with a two-car garage attached to one side. It was a brick house, bricks with hues of brown, and brown trim, and it was only a few years old. Phil and I liked it because there were stairs inside, and our bedroom on the second floor had a window that faced the court. Later on, we came to dislike the house because of its attached garage, which had to be kept clean. For some reason Saturday morning always dawned with an unclean garage and my father at the foot of the stairs waking us up in unkind tones. The garage, and the varied assortment of tools that Phil and I were said to have lost during the week, always left Dad in a perturbed state on Saturdays and I don't think we, as a family, have recovered from those mornings yet. I still can't walk into an unclean garage without my toes working back under my arch, a queerly guilty smile on my face, my fingers knotting explanations behind my back.

The room Phil and I shared in Saginaw faced the street, and by putting the headboards of our bed under the window we could stand on them and look out over our simmering world. During the long summer evenings this was a place of misery: daylight lasted until midnight it seemed, and we had to be in bed by eight. I'm surprised to this day that Phil and I don't have a grid of screen wire permanently impressed into our foreheads and noses.

My father built us desks, and on them we built model airplanes, plastic and balsa, conducted experiments in science, chemistry and biology, and drew sharp-nosed cars and planes. In between these consuming projects I stalked down the hall to our bathroom or to the other bedroom on the second floor, Sally's. She remained an infant for a period that I thought must have been embarrassing to her. Every time I walked down the hall to her room she was still a baby. I was sure that it hadn't taken me nearly as long to learn how to eat and speak. But Phil and I visited often, because she laughed at all of our jokes and was truly beautiful, and because she had a window in her room too, one with a different view. Her room was smaller than ours since a bathroom had been cut out of it, a bathroom with yet another window, but a window so high it was useless to us.

This window gave me the greatest scare of my young life. I was always sure, at bedtime, that Crazy Mike, or some other rock-wielding madman, lived in either the linen closet or the attic. I had to pass both these doors on the way to the bathroom before bed. I placed the greatest odds on the attic, since the madman would have to scrunch himself up into a two-foot-by-two-foot space to fit on a shelf in the linen closet—although this seemed plausible since I myself could do it. The short attic door stood at the head of the stairs and loomed up before you as you climbed the stairs, always growing, growing. It was a small attic, perhaps four feet high with slanting roof, and always unbearably warm. There were spider webs everywhere, spilled boxes of my parent's past and places my father said we'd fall through to hell if we stepped there. This seemed the finest abode of a madman I'd ever heard of. I'd step quietly out of our bedroom in my underwear, step quietly past the linen closet and then run for my life past the attic door, with its imperceptibly turning knob. On this night I arrived at the bathroom door winded but safe. It was closed. It was never closed. The light was out. That meant no one was inside. Or did it? I turned the knob slowly, the door jerked out of my hand and slammed against the sink, and then the pages of a book lying open on that high window sill were caught in the draft of open window and door and fluttered like hundreds of wings beating, and I turned and uttered the most blood-splattering scream I had in me. Perhaps this is why I write books today, attempting to deny and answer that old demon on the window sill, "I'm not afraid of books, I'm not afraid of books . . . "

Bacon came up the stairwell to our rooms, bacon and coffee and toast and oatmeal, and we descended with one hand on the rail and one fist in our eye. The kitchen and dining room were at the foot of the stairs, and my mom and dad, and often miracles: a dog, or Christmas, grandparents, or a box of pre-sweetened cereal, sometimes even little sausages that you could break up and shove into your oatmeal.

I don't think it's strange that I remember the downstairs of our house in terms of eating. Children are addicted to food in a way that goes beyond survival: while they eat they can listen, unsuspected, to the conversations of adults. In the kitchen we ate on stools at a bar my father built. My mother sat on the stool that was literally in the kitchen, Dad sat at the end of the bar, Sally next to him, and Phil and I on the long edge, our backs to the dining room and, somewhat painfully, the television. The Vietnam War was on TV while we ate, and after dinner the war was over. We had to carry our plates to the sink. We ate fried chicken and round steak and pot pies and pan fried potatoes that I crave at this moment. We had to eat great hordes of green beans with almond slivers in them. I remember them sitting on the outstretched tongue of my brother. He was trying to eat them while simultaneously holding them as far away from his face as possible. With my mouth full, leaning back on my stool against the garage door, I listened to my parents talk about their day and the people at IBM, and talk with us, Phil and Sally and me vying for their attention.

In the living room we ate popcorn late at night and on weekends, watching the Cowboys' miserable years, the triumphant reign of Matt Dillon over Dodge, and the inscrutable trials and visions of "Rawhide." In the living room we also ate our company dinners. Mom would push the couch against the wall and fold out her much prized Ethan Allen drop-leaf table, then cover it up with a slick plastic tablecloth from K-Mart to protect it from us. Whenever my mother's brother and sisters came to our house, we children had the peculiar pleasure of watching our aunts and uncle and even our mom revert to their childhoods. We ate in patches of silence, listening, watching, while June, Tommy, Melody and Linda sat at the adults' table and yelled and slapped at each other, laughing so hard they spit food across the room.

We listened. They knew their lives so well that they finished each other's sentences and spoke in a guttural shorthand language of half-sentences and broken syllables, names, years, facial expressions and body gestures, self-deprecations and insults. Their parents beamed among them and called them liars and cheats. Their husbands and wives huddled in front of the TV after eating and comforted one another, appalled and embarrassed by the changes in their spouses. We wanted to join in the fun but didn't quite know how. It would have been like sticking our fingers into a desk fan.

At the end of the living room, over the long table and my loud family, over the top of the rabbit-eared TV, a big aluminum picture window opened on the front yard and asphalt court beyond. And so, from the coolness of the house, I popped through the window as from one

paragraph to another. Our house was directly under the approach to Meacham Field. There was always a light plane overhead, or periodically a business jet, and we spent a great deal of our time looking up, trying to chase them on our bikes. It was the great hope of our massed childhood that a plane might crash on our street so we could pull the survivors out and then loot their aircraft.

We were lucky Saginaw was a suburb under construction. New subdivisions were being built all over the town, so there was a strong and ready source of fort-building materials. All three of our forts were on the south side; I suppose we expected some kind of invasion from Fort Worth. Building in the old style, we constructed our easternmost outpost of stone in a field used as a dump by a gravel company. My masonry wasn't any better then than it is now: although we explained the many niches in our rock walls as rifle ports, we couldn't explain the walls falling in on us, and so finally abandoned this post with scarred knuckles and bloody noses. Our headquarters were located three streets and a field west, a treehouse on the banks of Saginaw's deepest ditch. The tree, the largest in town, sprawled in a huge half-circle over a patch of black dust that flowed down the banks of the ditch and became black mud. With any luck the big treehouse would be empty when we got there. It was constantly being taken over by unknown, clod-throwing rivals, and our only hope for access was to get there before anyone else did and hold it for the day. When not gathering dirt clod ammunition, we were usually adding on, nailing steps up a limb to build a plywood cube and store clods there, or building a crow's nest as high as we could get it. Our materials were from the house builder's scrap piles: triangular, trapezoidal and half-round cuts of plywood, split two-by-fours, bent nails. Our tools came from I don't know where. No matter how hard we worked on our treehouse fort we couldn't seem to keep out the rain. We'd return to it after a storm and find our cache of dirt clods a pile of mud in the corner and our older friends' *Playboy* pinups streaked and wrinkled, which seemed to make them that much more valuable.

Closest to home, as close as my parents would allow, we kept a two-story plywood structure, floored, walled and roofed with old carpet remnants. It was in Eddie Kirkpatrick's back yard, across the court, and since his German shepherd also called it home its smell could sometimes be unpleasant. But when the dog would give it up and the last rain had dried, we used it for thousands of games of canasta and a single game of all-male strip poker that went no further than pants and socks because Mike's mom caught us. It was our handiest fort; we'd take most of our stolen snacks there and there refute, purge or sanctify each other's latest news. I always remember it as being hot inside, but I think that's because we rarely went back to this fort or any of the others after school began in the fall. During the winter the cats, dogs and birds held our forts as best they could, and generally did as well as we did in the summer, even to the point of improvements, adding a fine twig nest or a cache of hairless tennis balls.

School in the morning. And still I was surrounded by my family: my Aunt Linda was the elementary school librarian, Grandma was the head cook at the middle school cafeteria, and Uncle Tommy was a teacher at the high school. I felt a member of the aristocracy and at the same time under a great deal of pressure to prove I could read.

I sat my books down on the front porch, turned and pulled the front door closed with both hands. My next door neighbor, Bobby, balanced on his curb, waiting. We walked the two blocks to school banging our tin lunch boxes on our knees. On the way we passed the great maroon wreck, a smashed Bonneville in which a man was killed. His widow kept it in her front yard. My mother couldn't understand why she didn't get rid of it. Although we surreptitiously examined it many times, we found no trace of blood. We were pretty sure that the driveshaft that lay between the seats was the murder weapon, somehow rammed through the poor guy. Over the years high grass grew up around the car, the tires went flat and the paint faded to pink, and we knew there was a great sadness in the house behind. But banging our lunch boxes, high-stepping, we marched past, down the block, through the gate in the chain link fence and onto the school yard, a great flat expanse of sparse brown grass. If a kicked ball got past you it rolled for miles.

I was familiar with my teacher's beehive, my mother had one, but the row of letters above her helmet eluded me. I don't remember the exact moment of orgasm in my reading education. I mean, I don't recall the instantaneous transition from illiteracy, fingers in nose or mouth, to the realization that Spot was running, when I began to use my hands to turn the pages. But that first explosion draws me after it still. I'm still chasing Spot to find out where he's going. I think he's after a ball, and I mean to beat him to it and give it another throw.

It's a strange thing for a child to love books. They don't roll at all, and aren't thrown well, and their pages are already marked up. But they do stack in a neat way between both hands and your chin, and that's how I carried them from place to place. It didn't matter much what was between the covers; I was amazed with it. In second grade we had a reading contest: each title read earned you a rung up the ladder pinned to the cork bulletin board. Within a few weeks my ladder jumped the cork, climbed up the concrete block, bent ninety degrees and worked its way along the ceiling toward the light fixture. I'm afraid I must have been speed reading, because I don't now recall much of the contents. But my ladder rose to a great height out of the non-fiction sections: sciences, biographies, sports, aviation. Toward the end of the year I somehow fell down on the library carpet in front of the wrong bookcase and ended up taking home in my chin-to-waist stack of books a novel, *Where the Red Fern Grows*. This book was the first thing in my life that made me cry, outside of pain and a child's losses to an enigmatic world. And I resented it. I didn't go back to the fiction section for many months. I read instead *Thirty Seconds over Tokyo* and *Hiroshima* and studied *Airplanes of the World* as a bible. I didn't want to believe that something as terrible and unfair as your dog dying could really happen. The deaths of 300,000 Japanese hardly bothered me. It was a war, and war was something that had been agreed upon.

If I didn't walk home with Bobby, I went to a babysitter and waited for my mother to come pick Phil and Sally and me up. Mom owned a beauty supply (La-Von) for a time, which explained why my four-year-old sister had more wigs than most senior citizens' homes. Our sitter lived only a block away from Greene's Grocery, and on occasion I was sent there for bread or milk, a bit of freedom for the oldest. It was there, while lingering in the candy aisle, that I learned that Bobby had drowned that afternoon in our creek. It came over the radio at the front counter, and the women there asked me if I knew him, and I excitedly told them I did, that he lived just next door, and that he wasn't dead at all. I heard them say, "Poor little fellow," as I left. I didn't know if they were talking about Bobby or me. My mother picked us up later. Her eyes were red, and although she told us it was true, she wouldn't say anything else. She didn't know what had happened or what it meant or what would happen next. And this is how it still stands between us.

The only time I recall having been alone as a child was when I sat, late at night in our darkened home, in front of our lighted Christmas tree. Candy hung on it, strings of popcorn, strings of blinking lights. The tree stood in the living room, in front of the big window that looked out on the front yard, as close to the outside as we could get it. It seemed to me to be something beyond itself, something beyond even the idea of Christ and Christmas, something that held me with its blinking beauty and its loneliness. I always felt that as long as I sat under it I'd be protected, safe and loved. It was the closest thing, besides my family, to a God that I've ever had. A Christmas tree in a darkened house draws me still: my heart sighs, and I love it before my life; the world is all possibility and hope, and vaguely, somehow, dying. I haven't told you yet what it is. I keep saying, "somehow." I'm a missionary, trying to explain the text of the needles. "See this blinking bulb," I say. "It means your mother and father love each other, and this popcorn kernel, turned the way it is, means you'll only get half the spankings you're promised, and this limb, the way it sags, means your father can fix anything, and this brown needle means the world works as you will it, and this room with this tree, you alone in it thinking, means things are already changing, have already irrevocably become not the same."

20

My first experience with roofing was on our fort in Eddie Kirkpatrick's back yard. I didn't understand Newtonian physics at the time and started my shingles at the peak of the roof instead of along the base. Shingles laid in this manner funnel water into a house, rather than shed it off. And rather than risk this mistake again, Heather and I decide on a tin roof, which overlaps from side to side and requires a much greater degree of idiocy to mess up.

But immediately my basic character flaw rears its chagrined head. It's too big a job for me. Our roof is huge and steep, on a forty-five-degree pitch: too steep to even stand on, much less traverse. Mark tells me even a motorcycle cannot sustain a climb up this grade. We'll be forced to rig up some kind of outlandish scaffolding, and I can see myself dangling from the eaves, hanging by the claw of my hammer while my dogs bark and wag their tongues at me and Heather shuffles below holding out the headrest from the truck for me to fall on. I balk for a whole week with Mark and Robert's encouragement (Don's gone back to Michigan), calling professional roofers. I get hold of six of them on the phone, only four actually show up at the house, and all four promise to send a bid but never do. I suspected this might occur, as they walked around the house shaking their heads and holding their elbows. Apparently, no amount of money will get a professional roofer on my roof. To the wise this might suggest caution, but it simply pisses me off. I won't have a house without a roof on it. "We'll just do it ourselves," I tell the boys. "We'll start tomorrow morning." This is the last time I see Robert. He never returns to the scene of my bravery.

Mark does show up, though, and we agree he should work inside the house, on the third floor. In this way he can pass materials and tools out through the dormers to me on the roof as I need them, and be in a position to call an ambulance should I require it.

We begin by sheathing the roof with waferboard. That sentence says so much that it ought to be against the law. It's in the league of "They built a pyramid," or "We changed the course of the river." We start at the base of the roof with a row of two-foot-wide half-sheets; we're able to lean through the rafters and nail these sheets down. On top of these we then nail short lengths of two-by-fours so I'll have something to scotch my toe against as I lie on the waferboard and hammer it down. Mark works from the inside, shoving out a sheet and then nailing the top while I lie on it and begin to fasten the rest down to the rafters. I have never been so close to my lumber. Waferboard is a glued-up composite of cedar, spruce and who knows what else, can give you a mind-bending splinter under your fingernail and does smell like a wet horse's hide. Not altogether unpleasant. It's rather hard stuff, and nails ping and spring off it at an alarming angle to my face. Mark hands out another short two-by-four and with two big sixteen-penny nails I nail it on through the wafer and into the rafter. Approaching the peak of the roof with the sheathing, we're also building a ladder.

As I get more accustomed to my position in life, forty or fifty feet above the earth, my toe- and fingernails begin to retract, and I look about my perch. This isn't nearly as bad as I thought it would be.

I mean, so far. Mark's doing all the hard work in return for me risking my life. He has to cut almost every board and hand it out to me, and as we rise he must carry the lumber up a ten-foot stepladder in order to get it out. I, on the other hand, let the waferboard fall into place, strike home a few nails and have a few seconds of free time to wait for Mark again. He is rapidly coming to some uneasy thoughts about this arrangement but never suggests we rechoose sides. Every once in a while, to reinforce his decision, I say, "See Daddy fall off house? See Daddy bounce off tree and land on ground like chicken egg? See Mommy marry good-looking man with much bigger humpus and lots of money?"

Meanwhile, the big black vultures that continually drift and soar over our house are now much closer, so close that I can make out their eyes, which are interested. They're salivating with the possibilities, I suppose: rabbit on the roadside, human on a roof. And it's beginning to get hot up here. It's late May, and in Texas we're in the nineties. The sky is a continuously clear blue, and the sun, unfettered by cloud or tree, moment by moment seems to be focusing all its attention on me. I can't imagine what it will be like when I'm actually putting up the galvanized tin, which will reflect all that light and heat back up at me. I'll be an egg six inches from the skillet.

Mark talks another sheet up the ladder, coaxes it between the rafters and blasphemes the tree it was made from as it drops into place. Boy, it's hot. I start a nail, reach for a better handhold, drive the nail to its final resting place. I start a nail, wipe my hammer hand on my shirt, reach for a better handhold and drive, ping, the nail into outer space. It drops back out of firmament and ozone, through the buzzard-shredded entrails of jets, and bounces off the waferboard below, over the edge, passing among limbs and leaves to the upturned earth. The archeologist who excavated Thoreau's hut by the pond found a very high incidence of bent nails. I like to think I am in my league. I wonder if he ever smashed his thumb and for lack of a better curse yelled out, "Emerson!" It is just very hot out here. In a field, far below, across the total length of our 165 acres, Henry David Thoreau is standing at a fence. He steps on the bottom strand of barbed wire and lifts up on the middle strand so Emily Dickinson can step through without tearing her upheld skirt.

"Right, right," he says, eagerly, nodding, putting his long arm around her shoulders. They step crudely and daintily across the earth.

"Here," Mark says. I look down and take the cold Pepsi from him. "Don't think about falling," he adds. The Pepsi is achingly cold. Mark pushes up a half sheet, and Charles Darwin rises, brushes the dust from the seat of his trousers and leans out the open door of the boxcar. He looks down the long line of cars, mostly grain haulers on their way to Saginaw, then steps back inside, running his hand over his balding skull. Don Quixote, rested, squats against the far wall of the car, polishing the kneecaps on his suit of armor.

"You have, good sir, an interest in birds?" Quixote asks, his eyes flashing off his kneecaps.

"Yes."

"I go for the giant among the oaks. I understand his armor gleams as no other and that his feet sink deeply into the earth."

Darwin adjusts his collar.

"Fierce and unequal combat," Quixote mumbles. "Fly not, cowards and vile caitiffs; one knight alone attacks you!" Quixote says, a bit louder.

Darwin adjusts his collar, pursing his lips.

"So Leda," Quixote says, still looking into his brilliant kneecap.

"Yes?" Darwin says. "So Leda what?"

"Leda. A bird interest also."

"Why, yes. I suppose," he replies.

"Fear not," Don Quixote waves, waving closed the conversation. "I am experienced in adventures," and he begins to rub on his elbow guards.

Darwin, tilting his head in the manner of a bird, leans back into the wind and lets what hair he has blow.

I can hear children. Far away. At first I thought seagulls, but it's children. I look down through rafters, Superman vision, at Mark, who's marking up a piece of waferboard that will be half the roof on one of the four small dormers.

"Let's go to lunch," I say. "I'm hot."

He looks up at me, sweat-drenched and coated with sawdust. "What's a vile caitiff?" he asks.

We pick up Heather at the antique mall and make her sit in the middle, straddling the gearshift and getting equal doses of our two shirts. Mark is already shriveling at the mouth for a glass of iced tea. His mouth dries up to a little puckered asshole on his face. Seven times out of ten we go to Lou's Café in Azle, ten miles from our house, because she serves iced tea in thirty-two-ounce buckets. Mark empties his before lunch arrives. The waitresses have come to know him and usually leave a pitcher, but that's still only twice the size of his tumbler. When we get back to the house he'll have to stand by a tree so long he'll eventually lean against it. Our fare is simple and heavy, designed to put us to sleep rather than energize us. Heather and Mark usually dine Mexican, and I have a hamburger, ever onward to my grave, a fatty smile on my slippery lip. I tell Heather of our advances and retreats, she sympathizes but doesn't want to come out if I'm going to be on the roof. She tells me about the morning's disasters at the antique mall, and I throw my hands up in the air, and catch them again, but hardly anyone notices. Mark tells us what the twins have learned since the day before, some miracle of destruction or torture, or some remarkable insight they've made about him, their magnificent edifice of a father, king of spankings, instrument of unjust punishment. "Yesterday," he says, "they threw their spaghetti all over the kitchen, and some of it went all the way into the living room, and this morning I stepped on it in my bare feet." He beams. I look at Heather and try to figure out from the spacing of the freckles on her face why she wants four children. She is beaming too. I'm building a house that in the space of a few short years will be the scene of food atrocities. We pay our bill, moan our way out to the truck, and Mark, I swear to God, sleeps all the way back out to the farm, his teeth vibrating in his head. I want to fall asleep too, and suddenly hate Mark with a passion. I swerve hard around the last corner, slam on the brake.

"Armadillo," I explain to Mark, as he rubs his forehead.

"Did you hit it?" he asks.

"Yes," I say. "Yes."

It takes us several days to finish with the waferboard. The third story has been transformed into a great open barn, approximately twenty feet by forty-five, an expanse of plywood flooring marred only by the stairwell opening in the very center of the room. Heather throws all our scrap to one side and sweeps the whole place twice. I've never seen her as industrious. But it's a ruse. She finishes, pats the dust from her jeans, walks to a far end of the open room, turns, and dances in a twenty-by-forty-foot oval, pirouetting, glissading, leaping, jazzing and moderning and balleting all over the place. I look at her disapprovingly, but know our house has been blessed in the best way she knows how. She smiles, dips, bows, does a *Nutcracker* thing on the toes of her tennis shoes. She's a dancer. It's part of the reason I married her, hard thighs and amazing dexterity. She enjoys dancing too much, though. She can't keep a smile off her face while she's at it and refuses to burden her life with repetition. Her joy is the instantaneous choreography of a new movement in an old world. She finally wears herself out with it. It's hard to smile and dance at the same time. She wants to figure out some way to cover the stairwell so she'll have more room, and implies that my first job after breathing my next breath should be to construct her a barre and erect a wall of mirrors. Mark and I both sigh.

"Does your wife treat you this way?" I ask.

"Wait till yours has her first kid," he says. "You won't get any sex for a year."

Our tin roof comes in lengths of sixteen, ten and eight feet, mottled in that galvanized way, shards of blue, silver and grey. It's not that corrugated barn stuff, but an older type, with a lap on each side and an inverted-V ridge down the middle. Dad had to order it specially; even so, it's cheaper than a conventional asphalt shingle roof. Mark and Heather and I pick it up, examine it, talk it over, put it down.

"Yep, that's it," I say.

We examine it some more.

I say, "Well, we're wasting daylight."

Nobody moves. I don't have any ideas either.

"Well," I say, "we know it's got to go upstairs first. So that Mark can hand it out one of the dormer windows to me."

Heather and Mark jump on this. "Right, yeah, unh hunh," they affirm, and we all hedge for a couple more hours, lugging the razor-sharp tin up the two flights of stairs. Then, breathing heavily, we discuss possibilities, strategies. We don't own sophisticated mountain-climbing gear, and a rope tying my belt to a tree on the other side of the house would only keep me holding onto the rope all day—how could I then hold hammer, nail and sixteen feet of tin? My series of nailed-down two-by-four block footholds are great for traversing the waferboard, but I have to put tin on the waferboard, and removing the block, laying the tin and then nailing the block back down through the tin would sort of defeat its purpose.

We decide the only feasible alternative is a ladder, with a specially designed ninety-degree end to clamp over the peak of the roof. We spend the afternoon building this ladder. It is a twenty-eight-foot-long monstrosity of one- and two-by-four lumber that takes everything Mark and I have in us to get up on the roof. We lay our thirty-two-foot extension ladder against the side of the house and push our home-built ladder up it till it clears the eave. Then I race upstairs, climb out the dormer and down my two-by-four blocks till I can catch hold of it. I pull, one hand on my blocks, the other on the ladder, and Mark pushes, climbing the extension ladder. At the point where the eave of the roof becomes a fulcrum, Mark can no longer help. He jumps off his ladder and runs into a field, getting a significant distance between himself and the monstrosity in case I should drop it, which seems more probable than possible. At this point I can no longer move the ladder with just one arm. I have to plant both feet on a two-by-four, hoping its two nails hold, and pull for all I'm worth, pull the ladder up over my head. Then I take another step up. Repeat. Repeat. Repeat. Up the twenty-six-foot slope to the top of the house, and drop the catch of the ladder over. It falls into place, and now I simply have to move it another fifteen feet to the edge of the house so I can put on the first sheet of tin. I straddle the apex, pick up the ladder and begin to walk it to the end of the house. I realize as I walk that the tail end of this beast hasn't yet moved. Every five feet I have to creep down, pick up the bottom of the ladder and walk it over too. No wonder a professional roofer wouldn't take this on.

I don't know why my life is a nut in every bite. It must be so I'll have something to write about. I think shooting at charging rhinos would be simpler, more profitable and obviously less dangerous. If Hemingway hadn't already run from the bulls at Pamplona I'd go do that. But I don't think he ever built his own house, and so here I am, adding to the diversity of heroic literature.

From the mouth of a dormer issues a sixteen-foot tongue of tin, flexing and thundering in the slight breeze. I inch along the roof, take the tin gingerly from Mark and then quarter-inch back across the roof to the ladder. Carrying the flexing tin is like trying to control a pile of leaves in a windstorm. It does pretty much what it wants to and I just try to stay attached. But once I've laid it down on the waferboard it acts as if it's come home. I move it an inch this way, an inch that, trying to square it with the house. A tin roof with all its vertical lines, if laid out of square with the rest of the building, will make the house appear to be in the middle of falling down. The way a stick looks when it's half submerged in water. I finally get an "Okay" from Mark and put one foot on the ladder, the other on the tin, and then use a ring-shanked nail with a neoprene washer to tack it down. The incised rings on the shaft of the nail will grip the waferboard better than a conventional nail; the neoprene washer will seal out rain. And that's the whole point, to shed water. I can understand why many of our ancestors shunned water except to drink it. I'm just beginning to build my roof and already I'm disgusted with rain, its insidious rotting character, drip, drip, drip. A single leak can collapse an entire house in a matter of years. So I build my carapace of galvanized tin to keep out the weather. What God doesn't supply we must construct.

I skitter (I've become accustomed to the pitch and assume a reckless abandon) back to Mark for the next sheet of tin, fly it back to my ladder and fix it to my abode. Then with that row done, eave to ridge, I pull the top of the ladder two or three feet over, then the bottom, in order to start on the next row. I pull off my old two-by-four blocks, take tin from Mark, nail it down. Here I find, at a point where four sheets overlap, that one must use a pair of pliers to hold one's nail or one smashes one's thumb grievously. My thumb is all the more painful because in my present position, hanging onto ladder, hammer and tin, I can't jump up and down and thrash the air with my arm. I find I must take it like a man, and therefore let forth a string of obscenities I'll never be able to replicate, while I continue the job at hand. I am the sort who likes to throw himself face down upon the planet and beat it with his hands and feet when he feels any tinge of pain, remorse or regret. Epithets help but are by no means a substitute. Mark leans out of the dormer and says, "I thought you fell."

"No such luck," I say.

"We don't have a dainty tongue, do we?"

I let my hammer go, and it slides down the smooth tin, disappears over the edge.

Mark screams, then hangs his head in a sigh.

"I dropped my hammer, Mark," I say.

I hear him drop to his knees inside, sob, then gather himself together for the trip downstairs to retrieve my hammer. From the ground below he yells, "Drop everything else you're going to drop for the rest of the day so I don't have to make fifty more trips!" Then he mutters back up the stairs.

I move back over to the dormer for my hammer, but instead of Mark's curly head poking out of the hole, a pigeon's purple one appears. We frighten each other, and he flies away and I almost fall away. Two sets of these birds, marvels of fatness and flight, have taken up residence on the third floor. I have built a magnificent bird house, but pigeons are rotten housekeepers. They've built nests on the plywood of fiberglass insulation and shards of tin. Their eggs roll among this comfort and eventually find their way out, falling through one of the many remaining holes to the second floor or into the soffit. I can't understand how their species has survived with such lax habits. I mean, I'm not even a bird and I can build a better nest.

As predicted, my tin roof, with the sun doing what it does best, is roasting me alive. Mark hands out Pepsi after Pepsi, quart jar after quart jar of well water. The tin has actually become too hot to touch. Even the new sheet fresh out of the shade of the house becomes searing after only a few moments in the sun. The karate-chop portion of my nail-holding hand is red and tender from its contact with the roof. I lift my glasses, rub the sweat out of my eyes with the hem of my T-shirt.

"Boy, it's hot," I say.

James T. Feathers, the mestizo, and his good wife step hand in hand across the dust and stubble of a failed crop, a prone field, as if across the faded printed furrows of an old book. "This is something like the country I came from," he explains to her, sweeping through the sky his level and steady arm. "Very few trees on which to learn how to carve."

"It's such a long way from home," she says.

"No, this is home too."

"How much further?" she asks.

James T. Feathers stands, puts his hand up to shield his eyes and says, "There," and he points with the other hand to the barely visible tiptop of our skeletal tower.

"What?" Mark asks.

"What?" I ask.

"Maybe you ought to come in off that roof for a while," he says.

"Not yet, or we'll never finish in time."

"In time for what?"

I look at him in a blinding fog.

"I thought we weren't on a schedule," he says.

"We're not," I answer, "but I would like to get this roof on before Heather and I leave for Connecticut."

It's at this point that every thundercloud in the Western Hemisphere converges on our house. Not only these, for many more decide to actually form directly above us. For the next two weeks we are deluged with shower after storm, storm after thunderstorm. The pond stays full longer than we've ever known it to. During each rain the tin already on the roof steams, and after each rain I steam on the roof, steam literally, and steam figuratively over the lost hours and days. Our vacation is a fixed period. We have to be in North Carolina on June 6 to help celebrate Heather's grandparent's fiftieth wedding anniversary. We'll be gone for over three weeks. I imagine the soaking our house is getting now is nothing compared to what will happen while we're gone. Water falls on the waferboard, drips through the cracks, rolls down rafters and falls to the third floor, then to the second, to the first. I can feel the studs of the walls warping beneath me, the plywood of the floors separating and the very plates on the foundation rotting the teeth in my head. I lie in bed at night during a classic Texas three-hour drenching and imagine the great pools of water forming in my new house, the river running in the back door, through the kitchen, down the hallway to the foyer and out again through the front door.

I get up at six in the morning and meet Mark at the house, rain or not, and we sometimes piddle hours away waiting for an opportunity to get out on the roof. We finally get a break over a breezy span of three days when it rains only spittingly. My dad even comes down to help, which forces Mark onto the roof. He clings to me and the ladder and the house as if he were three months and we were breast and bottle. I keep slapping at him, but it does no good. It sprinkles while we work, eventually soaking us through, but we find the much more rain pleasing than the sun off the tin. The roof goes up quickly under these wet conditions. Mark works with the short pieces over a dormer and I work with the longer sheets on the main roof. Dad leans out the dormer window waiting for measurements and wincing at my every move.

I pick this timely moment to fall. It takes forever. I'm reaching out to the far edge of the tin, one foot on the ladder, bracing myself with the palm of my hand on the roof. As I lift my hammer the cohesion between palm and tin breaks, and I slide. Or roll is more like it. Sky, tin, sky and finally my hands catch on the ridge of roof over the library. I've fallen at one of the two places on the house where there is a possibility of stopping falling. I pull myself up and straddle the ridge. I'm shaking.

"Do you want to come in for a while?" my father asks.

I nod my head spastically and make my way over to his dormer, and he drags me in as to a lifeboat. Inside I check myself over for bruises, abrasions and concussions. My pant legs have a six-inch rip at the cuff. My heart is in an elevated state of being.

"You have to be careful, son," my dad says. "The whole house isn't worth a broken back. Take an extra minute and move the goddamn ladder over."

It's then I realize his heart is as raucous as mine. Watching someone fall is much more frightening than actually falling on your own. The worst part of a Sunday football game is the half-dozen replays of the quarterback breaking his leg. Because when you're falling you don't think about falling (apart from the first brief flash, "I'M FALLING!") but about not falling, about not dying. The survival instinct is very strong. Your life flashing before you, nostalgia, is bunk. The present moment is the only thing in the world that exists. All four appendages flail for a hold of some kind. Your mind accepts all possibilities. Yes, I can stand on the head of a nail, and yes, I can claw my way through sheet metal. I think that even if I'd gone so far as to slip over the edge of the house I'd have been looking for tree branches, passing airplanes, Heather's hand from a window and, finally, trampolines.

I stand inside, astonished that God would let me, of all people, fall. Dad and Mark are back at work, sawing angles through the tin. The roar of the backward blade through the metal, absolute pitch of excitement, metal screaming through its own kind, overwhelms even my bemused and simple ego. The world would go on without me, so I'd better catch up with it while I can. Besides, this noise hurts my ears. I climb back out on the roof, my toes almost bursting through the ends of my sneakers, and let my hands do what they do best: grasp. We finish in the late afternoon, nail on the last piece of tin and its corresponding section of ridge row. There's even time for some gratification. I take two antique lightning rods, with their round stained-glass globes and wind vanes, and attach one to the rear of the house and the other in the middle, where the ridge makes a little jump from one level to another. Then, at the front of the house, I plant a whale weathervane, in honor of Heather's New England heritage. I have to hold my compass way up in the sky, away from the magnetic pull of the metal roof, to accurately place the directional pointers of the vane. Then the four of us, Dad, Heather, Mark and I, all race downstairs and out to the rim of the pond and turn as one.

"Ya ha!" we say, again and again. Then the obligatory, "Thar she blows!" and at last the breeze changes a bit, from south to southwest, and the whale squalls around to that heading, and I am overcome. At last, at last after twenty-eight years, I have my own roof above my head. Let it rain, let it bucket, throw rocks at me; I'll be under my tin roof, secure, dry, civilized. Look at me: I'm reading Chaucer, and I have slippers on my feet. I shoot a rubber band at the feckless sky.

But there against the sky, to starboard, is only the spectre of a spire, mast without rigging. Our tower is still open to the ruthless, inscrutable blue. I'm too old to learn. My past accomplishments are all in the past. I sleep on the point of the tower all night, a bug on a pin, and in the morning I call more professional roofers. I call in steeple specialists, coppersmiths and unrequited lovers on the verge of suicide. They all say they'll do the job, put on two hundred square feet of roofing, for a mere $4,500, if I figure out a way for them to do it. I look longingly into the skies for a friendly helicopter.

We decide on a copper roof for the tower because we've heard it will last fifty years, just long enough for me to live and die so some other human sap can re-roof it. We begin, after extended sighs and much drumming of lips, with a trip to Dad's lumberyard for sixteen-foot two-by-eights and the longest nails we can buy. Most of the third story of the tower sits above the roof line of the rest of the house, enough that we can install five six-foot windows around it. After we remove the waferboard siding (two steps forward, one back) from the other three sides of our octagon tower, we finagle the two-by-eights through the windows and walls and nail them solidly just below the eave of the tower roof. Upon these we tack three-quarter-inch plywood and then attach a two-by-four handrail that wobbles in an unnerving way. Our scaffolding sticks out almost two feet all around the tower. We'll have to stand the sixteen-foot extension ladder on this and lean it on the tower in order to reach the very tip of the witch's hat. The copper comes only in three-by-ten-foot sheets, and comes freshly minted, bright as new pennies from the bank. We feel flagrant, lavish, working with it. We waferboard the tower first, then wrap it with tar paper, and then wrap it again in ordinary paper so the copper won't stick to the tar. I don't own any fancy brakes, metal bending machines or staple guns, so I have to fasten the copper with copper nails. Supposedly the nails will oxidize and form a waterproof bond with the sheet metal. I'm just hoping the extreme pitch of the roof won't even allow the water to poke around. The thin sheets of copper flap in the wind and make thunder.

My father, on a tractor in a field far below, points up at my tower, and I smile and wave at him. He continues to point. I smile more broadly and hold my piece of copper up in the blustering winds. He gets down off his tractor and steps over the clods toward me, still pointing. He's yelling too, but I can't hear him for the distance and wind. I guess he wants a closer look. I turn, glance up at the tattered paper on the peak, and there also, almost directly above the house, out of the north, is the blackest cloud I've ever seen in my life. It throws off arcs and bolts, and I realize that Lucifer is therein. I immediately climb down from that great conductor. Mark and I stand inside, waiting for the great crash of electricity finding no resistance. But it doesn't come. The storm cloud passes over leaving only a trail of wind, and Heather at the foot of the stairs, in tears.

I lay my hammer down.

"What?" I ask, knowing.

"It's Eliot," she says, breaking up, like a kid. "I found him under one of the buildings in the lumberyard. Bobby helped me get him out."

Both of the cats have been missing for two days and nights. We were pretty sure their absence was the result of their getting a flea bath at the local feed store. Mark, whose father is a veterinarian, says this shouldn't be done. Flea baths are fine for dogs, but cats incessantly clean themselves, and Eliot and Blossom probably ingested too much of the poison. I tell Heather it's not her fault, that the people at the feed store should have known, that she was trying to help the cats, not hurt them. This has been a bad summer for the animals, ticks and fleas all over them.

She has Eliot in the back of the pickup, on a towel. She's crying and crying over him—he was her favorite—and I don't know what to do. I've known her for three years, and I'm her husband, but I've seen her cry like this only once before: the time she told me how she felt as a kid when her parents divorced, but that was an old cry, one she'd had many times before, and all I had to do was hold her.

"We need to bury him," she says.

I nod, and reach down, pick Eliot up. He's still soft, and Heather scratches him as I carry him down the slope to the loose mound of earth behind our rock wall. We bury him there and pull a big flat rock off the top of the wall and over his grave, and I say, "Kitty," though it's Heather who suffers. She cries, and shakes, and cries, her arms loose at her hollow side, and although I hold her I am still the same stupid bastard without understanding, bright young novelist who cannot even console his own wife over the death of a cat.

She stops, finally, talking within her tears. "I haven't found Blossom. She must have gone off far away to die. I've looked everywhere." Then she gets back in our truck.

"Why don't you stay out here with us?" I ask.

"I'm all right," she says. "I just wish they didn't have to die. It wasn't their fault."

"It's nobody's fault."

"I'm all right."

Three days later Mark and I have almost finished. I stand on the sixteen-foot ladder and nail down the last small pieces of copper. The wind is strong and I hold onto the peak as often as I can. The ladder has a curious tendency to roll around corners. Dad is finishing his mowing in the field below, and Heather roams in the fields too, taking snapshots. The bugs and birds and chaff of the mowing have filled the air with a kind of rare smoke.

I am almost giddy with the beauty of our tower, the crown of our house. The copper, I know, will eventually turn brown, and then green, but now it's so shiny I can see myself in it. I have only rarely worked in precious materials before, but now I can feel the allure of the uncut diamond. I rub my hand across the smooth, shining sheet. The value in our turret roof is not only in its ability to shed rain for perhaps half a century, but also in and of itself; should we become desperate we need only sell our shelter by the pound: an investment of 488 pounds, 480 in sheet copper and 8 in nails. I am a magnate in metals. You figure it up. I bought at $1.40. We finish with three pounds of scrap. My hands smell like a pocketful of change.

I cap off the tower with a brass vase from India. I cut its base off, turn it upside down and *voilà*: Victorian tower ornament. But even with the final capping of the tower in place, Mark and I refuse to come down. We want the full effect, so begin ripping apart the scaffolding, the best bird's nest ever built. The handrail comes off first, then the flooring and finally the big two-by-eights themselves. We work late into the evening.

"That's the last time I'll be up there in my life," I say. "It's copper."

After we pull the last nail from the big lumber, we lope out into the grass under an almost full moon. We turn, and our mouths fall open.

"Look," we say, "look."

Our house is the moon. The tin returns all the light of the night to the blueness above, is radiant and almost unapproachable. Who could live in such a house? I'll have to lie out here in this open field, in a sleeping bag, just so I can exist in this light, under the influence of so much moon so near. "Look at that," Mark says.

One of our resident pigeons, coming from the direction of Dad's barn, alights on the ridge row, but only for a moment, and then is off again.

"I'll bet he took a dump," Mark says.

So we go home. Heather and I take the long drive, eight miles, back to our garage apartment slowly, and without many words, so silently that when we pull into our driveway and Blossom races through the beam of our headlights I can hear Heather's lungs swell, her heart go mad. It's been six days since we last saw her. We jump out of the doors so fast I forget to shut down the motor, and when I let go of the clutch the truck jumps and jerks to a halt. Heather picks up Blossom, saying kitty, kitty, kitty, over and over, and we take the cat inside, examine her under the lights and take her to the vet. She's all right, or will be after a few days' soft food and medicine, and Heather's very happy, grinning unstoppably, and so I grin too. But in bed later, she cries for Eliot again, sobbing already in that old way of being used to the thick ache of losing something forever.

21

I already knew, at age eleven, that the planet was a big place, but I had no idea it could contain something as faraway and unfamiliar as California. I'd lived in the state before, for a few months in Riverside when Dad was in the service, but all I could remember of that was the fall off our back-yard fence that broke my wrist, and the way tumbleweeds piled up against the barbed wire across the road. This California, I could tell immediately, was an alien. We moved to Cupertino (about two hours south of San Francisco) in October of 1970. Hippies, unknown in Fort Worth, moved through the streets here like amoebas. I asked for a hamburger at the one familiar landmark I could find, a McDonald's, and the cashier replied, "Why, shore, pardner, I'll get y'all a hamburger!" Everyone in the restaurant turned and looked at me. When I asked my mother why he spoke in such a way, she said I had an accent. But I couldn't hear it. Usually when I had something, I was aware of it. Everyone in the world knew me better than I knew myself. There was no telling how many appalling characteristics and mannerisms I had. Perhaps there were huge warts on my back too, at that place where I couldn't reach. Maybe my face was a wart and I just couldn't see it. Maybe I smelled like a skunk and was so used to myself I enjoyed it. It was in this state of mind I began my first day at the new school.

Our new house was Californian, a trilevel Spanish mission, multideck, sliding glass door, adobe, redwood, tile, wrought-iron-influenced shelter. It looked a great deal like the Alamo, the greatest fort of all time, and so Phil and I took it to heart as home and battleground, site of many spirited defenses of our Cheerios behind our arrow-ridden horse carcasses in the living room. It was a huge house, three thousand square feet, with enough bedrooms for each of us to have one of our own. I was moved to a bedroom off the living room, two floors down from the others. I liked the privacy, but the room was designed for an aged parent or a boarder and so had not only a bathroom of its own but also, horrifyingly enough, its own door to the outside world. I didn't sleep for a month, watching that door. I could be dragged out by a hippie and murdered and my parents would never know.

It was a big house that took up most of its lot. The front yard, landscaped to help sell the house, was a mere circle of grass with a cactus-and-rock border. There was no way in the world to go out for a long pass on it. The back yard was bigger, but it was a wasteland: every cupful of dirt in California that wouldn't grow anything had been brought here for fill. In the spring Dad bought a ten-foot-diameter pool, only three feet deep, and we thought it would be nice to set it in the ground, like a built-in. Dad, Phil and I hammered at that hole in the earth for weeks, with every pick, hammer and chisel we had, finally giving up with the pool a foot and a half below ground and a foot and a half above. For the twelve months we lived in Cupertino the back yard never grew a blade of grass, while all around us California bloomed. We poured a sidewalk around the side of the house, leading to the redwood deck outside the kitchen. Our initials—JC, PC, SC—should still be there. For a time we had a redwood lattice arbor above our deck, a bit of shade to have breakfast in. We spent three weekends constructing it, and one weekend tearing it down. One of our next-door neighbors complained to the city that it obstructed their view of our back yard, and since the city noticed we hadn't obtained the five-dollar permit to build it we had to demolish it. Dad could have applied for the permit and left the arbor standing, I suppose, but I think the idea of asking for somebody's permission to work on his own house annoyed him. He worked off the arbor episode by building (under cover of roof and curtained window) a huge entertainment center and bookshelf in the family room. Phil and I, we just chucked rocks and clods of infertile soil into our neighbor's back yard occasionally.

We lived on a court again, as in Saginaw, a soupspoon road invented with the subdivision, but this time we lived out on the handle instead of in the bowl. And it was just as well; Phil and I were ready to get out. We rode new ten-speed bikes to school. On the way home one afternoon I pushed the gearshift lever down too abruptly, breaking it off and ripping a deep gash in the base of my thumb. I cupped the blood in the palm of my hand to show it to my mom when I got home; I was very proud that I could bleed so much without swooning. But at home I found that Fleagle, our sweet hound, chasing a dog's life, had run across the street and was hit and killed by a car that never even slowed down. Phil and Sally and I sat in a row on the couch, and Dad kneeled before us, his hands moving along our knees as along a keyboard. He said, softly, don't cry, don't cry, I don't know, but if there is a heaven for dogs, that's where Fleagle is. I was already at the point where I understood he was trying to console me, instead of simply being consoled. I recalled that books of fiction had forewarned me. The cut on my thumb required three stitches. If we ever meet I'll proudly show you my scar, barely an inch long, proof that I'm alive.

On one of our last Saturday excursions from Cupertino, while we were still mourning Fleagle, we visited the Winchester House in San Jose. I was in a somber mood, or at least trying to be in one, but the house wouldn't leave me alone. The world takes little notice of the dead and even less of the mourning, and so the mourning are always looking for a subtle way out. Mine took the form of an honest exuberance for the first truly Victorian house I'd ever been in. This house had hallways. In every house I'd ever lived in, one room opened onto another, excepting small trots between bedrooms and baths. These weren't hallways as much as places to check your fly. Mrs. Winchester, bless her mind, thought spirits roamed her hallways and for many years built onto her house to accommodate them. She also had a firm belief that as long as she kept adding to the house she wouldn't die. These continuous additions and alterations produced a house worthy of ghosts: doors opened onto solid brick walls, stairs climbed an arthritic three inches per step, passageways had no exit. This, I thought, this, was a house: a house that lived a life of its own, with or without fleshy occupants, an independent and beautiful being, not just a shell to sleep and eat and pee in. What a wonderful and adventurous life could be lived here, what races could be had from room to room, how many books could be read and shelved here. From the high mansard-roofed tower you could see for miles and ages, you could sit there in adolescent solitude and sadness and pity your miserable life and your poor dead dog and your absolute incompatibility and awkwardness with the ubiquitous present moment. Jesus, it was a great house. Immortality through architecture. Mrs. Winchester, I'm thinking of you. It worked.

We moved in November of 1971 to Lafayette, a suburb of Oakland, across the bay from San Francisco, so Dad wouldn't have so far to drive to work. My life is just so many birds lighting on a telephone line. I keep on arriving but it doesn't mean I'm getting anywhere or that I have any idea of where I've come to.

So I go by the names. Lafayette. My parents managed to find us a house on another court, another street that was only an end in itself. We had yet to live on a street that actually went anywhere. I think my mother liked courts because there wasn't much traffic and my father liked them because they were close kin to the roads he grew up on, roads that led up into hollows in the mountains, one way in and one way out.

It was new territory and I was glad to get there, away from the spot on the road where Fleagle lay after he was hit.

For the first time in my life I wasn't living on flat ground. Cupertino was a sort of mud flat in a valley, and Saginaw was a flat place too, some of that valley between the Rockies and the Appalachians. Our new house backed up to a steep, grass-sloped hill that was topped with a grove of eucalyptus. Phil and Sally and I climbed it hours before we picked out our bedrooms in the house below. Our houses had always backed up to other houses; the greatest adventure of our old back yards had been a box below the fence and then a leap and hoist so we could peek into the next back yard. But this hill, and the open grassy hills beyond, couldn't be exhausted. We tried. The long grass became, took the place of, our rooms. We even slept there, carrying sleeping bags to the top of a hill, till whoever owned the range put cattle in and Phil and I woke in the middle of one night amidst hundreds of legs of beef.

"What now?" he asked.

"Get up slow," I said, remembering this line from the movies. We rose, draped our sleeping bags over our shoulders, and the herd parted for us in the darkness without a moo, as if they'd known us all their lives.

Our usual destination was Fossil Cliff, the hill two beyond our own that had been cleaved by God or nature to reveal sixty feet of sandstone mussels. We spent hours there picking over the broken fossils, clawing out a perfect specimen from the cliffside, scaling the strata of eons. I still feel the power and awe and satisfaction of reaching the summit of the cliff with two or three perfect five-million-year-old shells in my shirt pocket, turning and sitting there, legs dangling over the drop, seeing as far as I could see. Usually by the time we got home the shells were broken or crumbled to a fine powder. The fossils were only sandstone and would never last a lifetime. If they did make it home to our desks or bedside tables, they gradually turned themselves back into beach, and so my mother, much as Thoreau, threw them out the window because they required housekeeping.

Our new house was a complete departure from the old Spanish

mission: low and long, tucked into its leveled niche in the hillside. You could actually find one of your relatives in it. Phil and Sally and I each had our own bedroom again, and I find it a strong plausibility that Mom and Dad spent extra tens of thousands on a bigger home just to keep the peace among their children. The house was ten or fifteen years old when we bought it, and the previous owners almost stymied my father. They'd built in all the cabinets and bookshelves anyone could ever need, completely landscaped the front and back yards and even built a deck off the family room. It took my dad months to figure out something to do besides mow the grass. In conjunction with the state of California the siblings and I kept working at the idea of a pool. We knew if we presented it as added property value and building challenge my father could be coddled. In the spring he bit, and bought a large oval aboveground pool, set it among pines in the hillside and built a massive redwood staircase and deck around the rim. We knocked all the water out of it we could, climbing the hill, racing down, redwood for springboard and cannonball for quantity splash. Phil was especially good at this, his belly apparently immune to pain, splashing water halfway to the house, but we all bowed to my father: density and breadth and a kamikaze scream, he sent water cascading down the sliding glass door and my mother ducking behind it.

In the late spring of 1973, before we left on our next move to New Jersey, my father wanted us to see William Randolph Hearst's mansion at San Simeon. I was a child raised, in part, by television, and I was, by Hearst Castle, unimpressed. I'd seen it all before. The Beverly Hillbillies lived in just such a place. It was a great problem for my parents: every great edifice, historic sight and natural wonder they took us to was old news, the subject of some documentary or "Wild Kingdom" segment, or had been a movie set. I knew and was told that San Simeon had been built to impress, was built not as a home but as a place for other people to visit, built so that Clark Gable might come, that it was in fact a meeting hall, a hollow place. And here I was, marveling at the many pools, the ancient columns from Greece, the walk-in fireplace, one in a crowd of visitors following a tour guide, sucked in. I'd been in California for almost three years and had completely lost my accent. I couldn't walk through a dead man's house without being extremely self-conscious, my hands in the pockets of my windbreaker, hoping my hair would stay just so, and also extremely species-critical, unable to allow a lonely man a house that might draw friends. But what's amazing to me now is that I knew all this then, that while I was all boils on the surface I was calm beneath. I had the opinion then that my experience was no different from any experience since the clams on Fossil Cliff last burped. I was unable to let myself have a major hang-up because I thought hang-ups were all old hat and it was embarrassing to me to copy. I knew exactly what the process of adolescence was so I tried not to let it bother me. I knew the pimple on my nose would go away by the time I was twenty-five (it hasn't yet), and I knew I would be self-conscious about it till I was twenty-five (still am), and I was glad I knew, and hated the knowing. I have all my life been on the lookout for something original or something lost: they are one and the same.

22

Long before we sat down to design our house, we left our garage apartment looking for it, or at least parts of it, looking for the lost, the salvageable, looking for something that might be saved. Many houses will become our home. We went looking and still look for what might be, the puzzle piece under the couch, the key in the purse in the back of the closet, a glance of recognition between two old, old women. We travel with a camera, stealing the facades of houses, rolling slowly down the streets just off a courthouse square, memorizing trimwork and paint schemes.

"Look at that," Heather whispers, as if someone is pulling up behind us.

"But look at that," I say, and we wrench our necks around to catch a glimpse of a tower in the trees. The locals curse us because we drive so slowly, sometimes slamming on the brakes so we won't miss a Palladian window or a rusty weathervane. We pull up and pause in front of old houses like past owners, seeing what the new family has done. We're wistful about memories of the front porch and absolutely cannot believe they painted the window trim orange.

On a weekend trip a year ago to a Texas Institute of Letters meeting in San Antonio, we drove off the Interstate into the town of Buda and found our back door. It was leaning against a cream separator in the front yard of an antique shop and mooed at us as we drove by in our little truck. Cows grazed in the upper half of the door, etched into clear glass, and the rest of the door was painted several shades of white. It had come out of a dangerously leaning house in a dangerous neighborhood: the bottom of the door was rough-sawn almost three inches out of square, and there were three gaping holes above the doorknob for deadbolt locks. I guess it was the criminal's aesthetic sense that kept him from breaking the glass to unlock the door. The price was ridiculously cheap, so we didn't ask for a deal, and slid the door into the bed of the truck. The dealer told us of another shop just down the street that would have more old house parts. A hundred yards away we ran into the heart of Buda, eight or ten brick storefronts all on one side of the road. Inside Star Antiques old brass light fixtures hung from the close ceiling by the dozens. They were all polished and rewired and this is what sold me. The door would take hours to get back into usable shape, but the fixtures could go right in. We bought several hundred dollars' worth. The young couple who owned the shop were almost as unbelieving that we existed as we were that we'd found the fixtures. We went on to San Antonio then, and met with literary figures great and small, but in our hearts we would have rather been back in Buda, caressing old doors, holding brass fixtures above our heads to see them in a better light. From this point on I was done for writing. I wouldn't write another word till we began on the house itself. Writing wasn't nearly as thrilling as finding a porcelain doorknob.

And although we've since become seasoned in our search we're still under the spell of its weather, and leave our apartment this summer headed for North Carolina, Connecticut and Maine but looking for the fretwork in our gable. The road is something with which Heather and I are both familiar. It's almost a second home, so we leave the day after our roof is completed with a sense of relief and anticipation. In Texas you don't drive east until it's half past noon. We load our cube van in the early afternoon, letting the air conditioner run for ten minutes before we load ourselves. We generally use our van for local auctions and deliveries for the antique malls rather than long trips—it drains gas and is uncomfortable to drive—but we can't stand the idea of finding something along the way and having to leave it behind for lack of carrying space.

Unfortunately the van breaks down in Little Rock, Arkansas, and we're forced to fly to North Carolina while it's being fixed. While I'm calculating that our plane ride is costing us approximately five dollars a minute (eight pounds of nails or two drill bits), Heather is showing me what all the buttons above us do, asking what I'm going to have when they come by with the snack cart, yelling, "Oh boy, here we go."

"I wish we could fly over our house," I say.

"Wouldn't that be neat," she says, and stiffens, clutching my hand as we begin our roll down the runway.

Isn't it strange that I miss our house already. We'll be gone for three weeks. Mark will take some time off too but will come back early to do some odd jobs: brace up the front porch, start on the insulation. For a moment the insane thought that perhaps the whole house will be finished when I return passes through the holes of my bricklike brain. I hope a tornado doesn't come while we're gone. I mean, if one has to come I want to be there to believe it. I probably should have put more nails in the tower. I probably should have chained it down before we left.

In the evening we have a family dinner with Frances and Skipper. He toasts her, she him, and we the both of them and their fifty years together. Heather and I have been married a year and a half now, and much is made of the forty-eight and a half years we have to go till our golden anniversary. But I know that when the forty-eight years are over it will seem as no more than the blink of a day. The past is so close to the present that only a moment separates them. At the dinner table Skipper is twenty-five years old, three years younger than me, helping Frances wash dishes in their claw-foot bathtub, the only source of water in their first apartment. He eats her first meal, leaves her to go pilot a destroyer escort during

World War II, and she raises their first child for the first five years of her life. But he's written all this down before. I don't have to tell it again, once and years removed. We eat dinner, try to comprehend time actually passing, and then give up the effort for a funnier tale and the goodness of food, passing time as best we know how.

Sunday afternoon: in a great hall of the Carolina Inn of Chapel Hill, Heather and I stand at the end of a receiving line greeting septuagenarian after septuagenarian. They're all friends of Frances and Skipper and have gathered to help celebrate this coming of age. Heather and I try to talk to them about her grandparents, but her grandparents have already talked to them about us. The ladies, North Carolina beauties, hold my hand in both of theirs and congratulate me on my books, but then they get down to the nitty-gritty, sincere appreciation burning in their eyes: is it true, they ask us, that we're building our own house? Writing accomplishments are old hat to them, or old pen: Skipper (Vermont Connecticut Royster) has been a professional writer for fifty years or more, culminating with a stint as Editor-in-Chief of *The Wall Street Journal* and a couple of Pulitzer Prizes. What these ladies and their fewer husbands are impressed with is hammer to nail, handsaw to board, pencil to graph paper. Their interest is polite but also intense. It's as if they're all starting a house next week. In the shelter of the Carolina Inn we stand next to a glistening swan ice sculpture and tell the best friends of Heather's grandparents about the house we plan to live in for the next forty-eight and a half years. After a while, realizing that these people love Frances and Skipper, we understand that they'd be interested if we were building stone boats, but we go on telling them about our house, enamored of their eyes and our subject, the sound of our own rapture. Skipper joins our group and listens for a bit, beaming at Heather, sure as every grandfather is that his grandchild is the last hope and salvation of the species. At a lull in the conversation he reminds the assemblage that this work shouldn't interfere with my writing. Skipper has the appalling and somewhat undermining ability to see the realm from all perspectives at once.

"But," I say, "I have to finish the house to finish the book about it."

"Nail faster," Skipper admonishes, "because it's the fourth book that makes the writer." He's also made this statement about second and third books. He smiles broadly, pivots on his cane and is off to the next gaggle.

In Little Rock, Monday midday, we rent a car and drive to the Ford garage. The mechanics assure us the van will be ready by Tuesday at lunch. At present the engine is spread across the floor of two whole work bays. We sigh, check into a hotel, spread open the yellow pages on the bed. We cross-reference all the antique shops, circling streets on our map. We'll have the afternoon and the next morning to travel to them all.

And we manage to make some finds. We pick up a couple of odd doorknobs at a mall out on the loop. They're brass. One is so tiny it could almost be a cabinet pull; the other isn't a knob in the usual shape but a sort of flattened spherical target. Both knobs are worn, nicked, have earned their patina, and most importantly we don't have any like them. For some reason we've vowed to make every doorknob in the house different. Any blind person will be able to tell what room they're entering by the grace of its doorknob. Each knob will become a familiar hand by turn. It's odd to think I may come to associate each of my children with a particular door latch, that the feel of brass, or glass, or porcelain, will riddle me with my child as I enter their room.

In a warehouse just off the Arkansas River, a can's kick from the closed Victorian railroad depot, we stumble upon an architectural mine: three floors of old windows, doors, fretwork and even entire interior rooms. Several items in this building are priced at just under what we plan to spend on our whole house. We wander through it all, the only customers in the entire place, and are finally overcome with the sheer quantity of merchandise. If it had just been a couple of rooms we might have bought it out, but we're intimidated by so many doors. After flipping through a row of twenty we decide we'd better concentrate on only the truly magnificent. On the third floor I find a pair of huge exterior corner brackets, quarter-circle art nouveau suns. They're made of pine and have eleven coats of paint minimum, but that's fine; I'm going to make it twelve. I carry them downstairs to the counter one at a time. On the first floor Heather rushes me with a snake-in-her-sleeping-bag whisper.

"Come here, come here!" she whisper-screams, grabbing my wrist and jerking.

Behind a pile of green shutters she points to a toilet.

"Isn't it beautiful?" she asks.

Although I do think it is beautiful beyond words what I say is, "How much is it?"

"There's no tag on it," she says. "You go ask them."

We both walk up to the sales counter. Heather hides behind me to conceal her anxiety. She wants this toilet bad.

"How much is the old toilet?" I ask, pointing.

The old man and woman behind the counter look at each other and smile. "Isn't it beautiful," she says. "That toilet is from the Kellogg summer mansion on Lake Michigan. The Kellogg cereal people. That's why it's in such good condition. They only used it a few weeks out of the year."

I know the toilet is from England, it has an English registry number on it, but it's possible that somebody like the Kelloggs could have imported their toilets.

"We've only had it about a week," they say. "We had a blue one just like it but it's already been sold."

I'm still wondering how much the toilet is.

"It's a hundred and fifty dollars."

"We'll take it!" Heather screams from behind my back.

"Wait, wait," I say, trying to be as cool as porcelain. "It's got a weird hookup, honey. It's just a bowl. I'll have to find or make a tank and all the hardware."

"You can do that, can't you?" she asks.

If she hadn't added the "can't you?" I might have argued on for a couple more sentences, but my supreme handyman ego forces me to reply, "Sure."

"We'll take it!" she screams again, sure that the customer who's just entered the door is looking for a brown transfer English toilet bowl. And it is as pretty as something like a toilet can be. The porcelain is covered with carnations or tulips or something on the outside, and the inside is stamped "Waterfall Closet." We'll put our cereal bowl downstairs in the guest bathroom so we can tell of its heritage many times over.

We pay for our prizes, arrange to pick them up later in the cube van. But on the way out I spot another pair of brackets hanging from the ceiling. These are truly huge, perhaps four feet by four feet, with a radiant Victorian sun emerging on both sides. The old couple explain they're from a house in Toronto, Canada.

"We'll take them," I say. "They're supporting our front porch already, just on either side of the front door," I tell Heather, and we leave, touching everything as we go.

In the evening, after all the shops have closed, we throttle down and enter a grey-shaded area on our map titled "Historic District." It's a fantastic cache of old house parts, still attached to their old houses. We rumble by the mansions, sneaking views. A few of the houses have been completely restored with brass fixtures gleaming from every window. We are almost as proud of them as their owners must be, but what catches our breath – we are fish on the bank – are the still-derelict, vine-crept, completely unhaunted houses that are the air outside our open mouths. It's not that we want to pick them over, tear mantle from flue, although we understand that urge too. We want to stop the truck at one, climb over the wrought-iron gate, wade and stoop through the brush and scrape the old paint off the porch posts. It's a house, and lives once rose and fell there, and still could, in a magnificence that we can only dare to imagine.

"If only we could have this house on our farm," Heather sighs, wistful as the South.

"No," I say. "Not in a million years. Maybe in a hundred years, because our house will look like this then. But we don't want it now. We might be living in our house in six months. Restoring this one by ourselves would take ten years. Can you imagine the exasperation of rewiring and re-plumbing it?"

"But look at it," she says, never looking back at me.

"It's like a rusted out basket case of a car in a field," I say. "It's a great discovery, but it would cost more than it's worth to get it back on the highway."

She spins. "You take that back."

"I knew I didn't believe it as I said it," I apologize, and sigh with the old house myself. I put my foot back on the accelerator, as softly as possible, so Heather won't notice. "We've got to save what we can," I say stoically, and we pull away, bemoaning fate: only one life to be led, so many houses, and so little time.

The next morning we visit a few more shops and then go to pick up the van. The mechanic slams the hood as we walk up and says, "Good as new." The truck has only 29,000 miles on it, so his statement doesn't impress us. We give him ten days of our time for two days of his, and he and his friends let us go.

Somewhere around Brinkley, Arkansas, I look up out of the roaring of my skull and see the contrails of a B-36 heading southwest. I see the contrails, connecting two clouds, but not the airplane, which I know couldn't be up there. There are only three B-36s left, and none of them has the remotest chance of becoming airborne. But no other airplane that I know of has ten engines to leave trails. Until the Boeing 747, the B-36 was the biggest airplane in the world. It never fired a shot. They called it the Peacemaker.

Heather screams, "What was THAT?"

I look over, purple lines all over the place because of the contrails in my eyes, and say, "I think it was a B-36."

"You're driving over the road titties," she screams.

At Yarbro's Antique Mall in Jackson, Tennessee, we find a doorbell button that will work with a bell we already have.

When it comes time we'll put our house together like a Frankenstein's monster, each part with a previous owner and a tale to tell. The brain will be a mix of mine and Heather's, as diseased as they are. We've made this trip twice before, and countless shorter journeys in and around Fort Worth, and even a jaunt to Kansas City once, looking for what might become our house. Our doorbell will go on a set of doors we bought in Austin, Texas, but which originally came from a house in England. Etched curtains hang in the glass of the two doors and transom, and below them peacocks strut on a verandah that overlooks a Middle Eastern town and bay. Our house will have a cosmopolitan impact. The guest bath just inside these front doors will have a stained-glass window that supposedly came from Scotland. It's small, perhaps twelve inches by thirty, and has a circular panel in the middle on which someone has painted a rather forlorn retriever. Heather's famous toilet will go in this room too, along with the tiny clawfoot tub we found on Fort Worth's south side at Hearne's Wrecking Yard.

We bought seven oak office doors from Lindy Hearne also, all three feet wide, most of them painted a rancid green but a few still showing their tiger oak grain and one even retaining a hand-painted "PRIVATE" across its face. We figure it went either to the president's office or to the toilet. I'm going to use the door for the library, so that when Heather barges through it I can feel violated. After we'd picked out another tub at the wrecking yard and watched, for fifteen minutes, four old men pulling nails out of older lumber, we went inside to pay out. It was then that Lindy told us about the fretwork. He had a large piece on consignment at an antique shop downtown. He'd had it for a while, trying to sell it for $2500, but wanted cash now and was willing to sell for a thousand. We went to the shop unbelieving, but were caught up in the fervor and donated a thousand dollars to our faith. It was and is a magnificent piece of work, twelve feet wide and ten feet tall, designed for an archway between two rooms. The main display, a spray of stick and ball and sawn work, is supported by a column and a half on each side. Lindy said it came from an Arkansas house and that when his father bought it, it was painted thickly white. They hired an old woman to take it all apart, strip off the paint, put it back together and give it three coats of tung oil. Unbelievable. We plan to use it at the foot of the main staircase, between the foyer and the living room. I can hardly wait to put it up.

At an auction of Netherlands antiques on Fort Worth's east side we bought an etched-glass door depicting an almost life-size baker with traditional hat, holding a steaming loaf of bread and a knife. This will be our pantry door. At a later auction of antiques from Holland we were startled to find a door with an entwined H and J etched into the glass. It made us wonder if there were Howard Johnsons in Amsterdam.

In Kansas City we found a massive cast-iron pedestal sink. The price was outrageous but still less than half of what a new one would cost, so we brought it back to the garage.

Even before Heather and I were married we retrieved from Kentucky in our little Dodge pickup a fine double-columned oak fireplace mantle. We weren't thinking about a house then but liked it for its beauty as a piece of furniture.

At a Brimfield, Massachusetts, antique fair a summer ago we found a nickel-plated shower fixture with a curtain ring, a piece of doorway fretwork, and a set of tin roof endcaps that reminded me of the prow of a Roman warship.

Just a few months later at an estate sale in Fort Worth we paid two dollars for a five-paneled, four-foot-wide pine pocket door, complete with rolling hardware. There were some thirty dirtdobber nests attached to one side that went along with it. From various shops in Dallas we've brought home a set of five stained-glass windows with red hearts set into their centers, a fine old worn library ladder and a huge beveled hall mirror, framed in rope carved oak, that we'll put on the stair landing. In Denton, an hour north, we paid another two dollars for a painted butler call, an Edwardian electrical device which uses bells and flags to notify the servants that they are summoned. Heather thinks it will be handy when one runs out of toilet paper. And in our own antique malls in Azle and Burleson we've found doorknobs, closet hooks and other hardware and a wonderful etched and beveled window announcing, "Home Sweet Home." Our garage, where all of this treasure rests in an attic-like atmosphere, is my favorite place to rest. I go in, pull down the overhead door, put my hand on a doorknob and envision the future, a future made up of all the past around and behind me. I imagine the parts as a whole and the whole as more than the sum of the parts. I imagine a seamless Frankenstein, a great hollow beast of a home, a beauty on the plain, tribute to my fevered genius.

After Tennessee we spend most of the next two days bumping along America's highways in our van, bound for Connecticut. We're days behind, and Heather is anxious to see her family. We stop solely to pee and gas up, and occasionally to stretch our legs at an antique shop. Heather reads as I drive and looks up only when I say, "Look at that," pointing out the same old house that fifty million other Americans pointed out this year as they drove this stretch of highway. Heather goes back to her book, reading through any turbulence, through the West Virginia Appalachians, around curves that would surely empty my stomach. She is an acrobatic reader. I drive along, spitting out sunflower seeds till my tongue bleeds, sipping cold Pepsi, writing the next chapter.

It's the middle of the afternoon when we arrive and no one is home, so Heather takes the opportunity to pull out of the van a rather ornate Victorian fainting couch that we bought in Kentucky and put it under a front-yard shade tree. She waits there, reading a torrid novel. I suppose this scene is partially congruent: her WSM (Wicked Stepmother) and dad's house is a late Victorian model they claim came out of the Sears catalog. It overlooks a small cove off the Niantic River. There is more green grass on its quarter-acre lot than on our entire farm back home.

We enter their house through the portal of privilege, the back door. Their back door, as most back doors, opens into the kitchen, where the most secrets are kept. Front parlors and foyers are as dead and lonely as an empty bottle on the side of the road, but kitchens are always at least as fresh as the morning's cold coffee, hard biscuit, stain of egg. You know the species was still surviving as little as a few hours ago. A kitchen is always expecting, rather than waiting. Beverly's kitchen acts as if it thought we'd never get here. Heather wants our whole house to be like this kitchen—this kitchen, and the re-covered sofa in the living room, and the almost odorless potpourri pillow on the sofa. She recognizes the kitchen as kitchen of her childhood, even though it resided in several different houses across the country. Beverly's pine drop-leaf is here, her open cupboard, her dishes and hot pads. I know that there are Apple Jacks in the bottom of the cupboard because Beverly knows I like them.

Heather flops on the sofa, pulls the potpourri pillow up under her nostrils, and although we are two thousand miles from home she is home. This old Sears house has been settled. It is so settled with a family that I am surprised it hasn't fallen into its own basement. Heather and I drop into it like an old cotton shirt wadded up and thrown into the corner. We are so at home that we don't even notice ourselves. I am picking stuff out of my nose and putting it between the couch cushions before I can catch myself.

I redeem myself by carrying my own hammer and nail apron. Rob and Beverly have either partially or completely remodeled every house they've lived in, and this house is no exception. They're turning the third-floor attic into another bedroom. They've already rewired and insulated, and I help with putting up the sheetrock. I can tell Rob is impressed when I take the hammer and my father's lumberyard-logo nail aprons out of the back of the van. It's not every guy who carries a claw hammer with him everywhere he goes.

On the way up to Maine to visit Heather's grandmother we make a small side trip to see Walden Pond. It startles me. It's so big. And there are so many people on its banks. And then I realize I don't see it at all, that it's at home, on my bookshelf.

Grammy Hutton lives on the Piscataqua River just up from Portsmouth, New Hampshire, on the Maine side, in the town of Eliot. Rob grew up here, and so did Heather during each of her summers. It is the one spot on the planet that wrenches her heart. The old house, built circa 1830, sits on a bank above the tide-racked river. The house is old, and Eliot is older, and Heather's Aunt Corrine lives within casting distance up the river in a Cape Cod cottage that's even older, but the river oozes with age; it's so old it hasn't even bothered to notice us. It flows out to sea and the tide brings it back in. Heather and I, first thing, take ourselves down to the rock shore, say hello by shaking our hands in the cold, clear water. Then we look for shards. Grammy's piece of river shore was once the town coal wharf and apparently a good spot from which to sail dishes into the Piscataqua. Heather has bags and bags of broken dishes she's collected over the summers, a girl after my own heart. We wait till the tide is at its lowest and then venture out into the rock, muck and seaweed in our sneakers. The shards—creamware, majolica, ironstone, flow blue, crockery, spongeware—date as far back as the eighteenth century and as late as the 1950s: we find many pieces of Depression glass and Fiesta ware. We make our past irrefutable with our trash, our record on the planet. It may be the only thing that lasts, trash, the only thing we despise enough to throw away. The last words of the species will be on a broken cup on the bottom of the Piscataqua, something like "Shenango China" or "microwave safe" or, if we're lucky, "World's Best Grandpa." We hoard our shards like pieces of eight, locking them in a shoebox and labeling them "Summer of '87, Eliot, Maine." We have great plans to have great plans for them.

On our last Sunday in New England, Heather and her sisters plan to spend the day at the malls and beach, so I take the van and drive three hours to an auction in Pittsfield, Massachusetts. An advertisement described the auction as consisting mainly of parts, a shop cleaning of a longtime antique furniture dealer. The ad cautioned that there would be no complete pieces. I knew immediately that some part of my house was sitting in a box in Pittsfield, Mass. The antique dealer saved anything salvageable off of otherwise ruined Victorian furniture: carved oak crests off of dressers and organs, gargoyle and lion heads from buffets, finials from beds, and boxes and boxes of applied carvings: leaves, grapes, cherubs and even flowerpots. I buy to the degree that people start turning around and sneering at me. I'm sure they think I have the missing bottoms, tops and sides to these parts. I want to stand up in the middle of the auction and explain, "I'm going to use these finials on my fretwork, and these crests, they're going above each of my doorways, and these lions, I'm going to use them for newel posts." I could go on and on. Instead I sit here and wave my bid card, paying a nickel on the dollar for my house and smiling stupidly.

Back in Waterford, Heather and I pack the van, placing the fainting couch up front so Cutter, her little brother, can use it on the long drive home. We'll put him in charge of the cooler and snack sack. As we're packing, Beverly comes out and gives us a cast-iron Victorian hat hook.

"For your house," she says.

So the evening before we leave I let her read the first few pages of this book. I'm a little queasy about giving them to her, since the part about me and Heather fooling around on the house site is in them.

We've been away from home so long we're almost lost. Heather's happy because Cutter's going home with us, and I'm happy because for the first time she's not crying as we pull away. The only one crying is Beverly, waving us off from the front yard. She'll go back inside now and fix Rob lobster rolls from the leftovers of the feast we had last night. He'll come home at lunch and eat them, and they'll talk for a moment about Cutter, what a long trip he has before him, and maybe they'll even figure the hours and mileage and guess we're almost to Port Jervis by now.

Halfway home we pull tiredly into Austin, Indiana, to spend the afternoon and night with my Grandma Coomer. The yard is trim and neat. She hires a boy to cut the grass now. I knock on the glass storm door, but no one comes, so I step inside and find Grandma wide awake, watching TV in the front room. She doesn't have her hearing aid in. "GRANDMA!" I yell, and she turns, smiling before she finishes her turn, her beautiful smile lifting her eyes to mine.

"I didn't hear you come in," she tells me, the cool raspiness of her voice washing over me as we hug. She hugs Heather and Cutter next, and says to him, "What a strange name for a little boy."

We go to Kentucky Fried Chicken in Scottsburg for lunch, and then Grandma, sitting in the passenger seat of the big van, and Heather and Cutter on the fainting couch in back, guides me to the cemetery where Grandpa is buried. She wants me to see his tombstone. I park the van alongside the chain-link fence and we all walk into the graveyard. She leads us up a slight incline to a brown tombstone with my name on it, and we all stand in front of it for a while. She doesn't cry. She's been here many times. "I've been thinking," she says, "about getting two vases for Joe's tombstone. They're made of stone, too. They cost sixty dollars, but they don't turn over."

"I think you should get them," I say.

"I want to," she says. Then she says, "He's pretty close to this tree."

We go back to Grandma's house and watch TV and talk into the evening till we all go to bed. The Austin of my youth is still with me but doesn't seem to be as noisy. I listen for squalling tires, yells, gunshots, but there are none, and I fall asleep long after Heather does. I listen to the house calling my grandfather's name endlessly. Grandma is still up in her room, reading, lying in bed with the blue wash of the TV folded over her. In the morning she cooks us breakfast while we load our suitcases into the van. We're getting an early start. As she cooks and I make trips back and forth through the kitchen, I steal glances at her and try to see her as my father would, younger, with a son's love, but I can only see her as I will, her face crinkled softly, her eyes as witty as her humor, her insight sharper than almost anyone's I know. I want to take her with me, but she won't go.

I get in the van, roll down the window and wave at her on the porch as we pull away. I hug her as long as she will let me, and then I look at my wife.

In Lancaster, Kentucky, on our final leg home, we find the antique that somehow, sadly, completes our search at least for this trip. We drive the remaining nine hundred miles home lost in its mystery. The ladies at the Lancaster Antique Market tell us they don't have any idea where it came from or what it means. Heather, not trusting the back of our van, holds it in her lap almost all the way home. And it's just a stone, a slab of worn white marble two inches thick, eight inches long and six inches wide. There's an inscription engraved into the stone, in script and block letters. We still aren't sure of its significance to the people who had it made, or even what it means to us, a young couple building a house 110 years later. But we keep running our hands over the stone, and running our hands over the stone, knowing something's there for us, hands over stone, some faith or hope, some pervasiveness. The epitaph reads:

We cherish thy memory
our
HOME
Dec. 25, 1877

23

New Jersey is the closest I've ever lived to Walden Pond, and the furthest away. It was where I learned to become a social being, a link in the chain, a brick in the wall. I learned how to blend. No one could have picked me out of a lineup. I know this is so because I distinctly remember choosing my wardrobe. My two years in Ringwood were the only years in my life that I haven't worn blue jeans. I wore plaids. Plaid slacks, shirts and even argyle socks. I'm not sure but I may have had a white belt. I wore these things in order to wash into my surroundings: everyone else wore these things. I knew my disguise worked, because in crowded hallways between classes upperclassmen ran into me.

We flew into LaGuardia on a 747 jumbo jet, approved of my

father's house selection, flew the five hours back to California and got in the car for the trip back to New Jersey. Zeke, our hundred-pound black Labrador, lay sprawled on the floor of the rear seat, his slobber towel under his chin for the whole five days of the drive. Phil and Sally and I rested our feet on him. Dusty, a toy poodle prone to carsickness, wedged himself between my mother's door and the seat and popped out of the car like a spring when we stopped for gas. We were on the road again; we knew by the dog food in the trunk that we were in that limbo of the American highway, between jobs, schools, houses and even friends, with no other responsibility than getting from one point to another. We asked questions from the back seat about the future, questions our parents couldn't possibly answer. So they answered in generalities: "Soon," "Sure," "Of course," "Not now," "Shut up." I leaned over the front seat and snapped a photograph of the road before us, turned around and snapped one of the road behind us and then finished out the roll snapping my fingers at Zeke.

Fifty-one Dewey Drive had been cut out of the foothills of the Ramapo Mountains. Our house, a two-story Colonial with attached garage, sat on a curving subdivision street alongside other Colonials with attached garages. It was a novelty to live on a real street; it seemed unusual not to have to leave your house by the same route you came on. I could go out the front door and actually choose a direction. Our front yard was a large, grassy tract that sloped steeply to the street. Sally used it as training grounds for her pony, Tony. The back yard was a pine forest until it ran into the back yards of the houses on the street above us. Tony's pen and stable were there. His odor swept down the hill, lingered at our back door and entered the house with us. Tony's smell mingled with the freshness of Pop-tarts in the toaster oven, and Zeke's almost constant diarrhea.

Zeke didn't get to stay in New Jersey long. I think it was his choice. He must have missed California's range and disapproved of New Jersey in some canine way. He had to live on a chain most of the time, and to show us his desperateness he did great runny streams of diarrhea on my mother's royal-red parlor carpet when he could. He had diarrhea because he held it in till he could get inside to the parlor. My mom would kick the door open with groceries in both hands and Zeke, whimpering, slammed past her on his way to the parlor. She'd drop the groceries then, pick up a broom or mop on her charge after him, and together she and Zeke would make intricate patterns on the red carpet, circles and figure eights of a light brown liquid that Zeke parted with as quickly as he could. Mom cured him at last, broom to ass, and Zeke, out of boredom, frustration, arrogance, took to breaking his chain and roaming the streets.

He came to know a great many people in the neighborhood, some by rocks and some by petting, but was finally forced out of the village by a fellow canine. A phone call on a Saturday morning informed us Zeke had killed another dog, a poodle, on the next street over. We had no doubt that Zeke had the power to kill a dog, but we never thought he would use it. We'd seen him pin several dogs, but he never carried his weight out, especially on something like a poodle. Within a week Dad had carried Zeke and Tony to Indiana to my Uncle Thomas's farm. The authorities said it was either this or put Zeke to sleep. Zeke lived in Indiana for another few months till we took him to Texas to live with my grandparents. Months later we learned that the poodle had been a hand taller than Zeke, that they'd met in the middle of the street, that Zeke bit him once, in the back, breaking it.

I felt great guilt over the dead dog, especially since I thought they'd be burying it in a shoe box. The family who owned him had a daughter in my class at high school, and she didn't speak to me for the whole two years I was there. We passed silently in hallways, the words "DEAD DOG" thronging both our brains till we rattled like locker doors. She was very popular, a cheerleader and student council officer, and my dog killed her dog, and so I skulked the school grounds as if I'd done the deed myself, an old black Labrador from California who killed defenseless New Jersey poodles. At home it was a bad time for us all, losing Zeke to Indiana, but we were glad he'd won the fight and knew, with Dad's job, that we'd be moving again someday, to a new neighborhood where they wouldn't know Zeke's reputation and he could live with us again.

So I spent the majority of my first year in Ringwood underground. Our new house had a full basement and I lived there after school. I hit tennis balls against the concrete-block wall till I was exhausted by my flawless opponent. I could never win a point and constantly chastised myself for it, wiping the sweat out of my eyes. The harder I played the more mistakes I made. I played till I was so mad at myself that I quit and took my ball home.

On the far side of the basement I worked with my hands on top of an old bench. There I refinished a dovetailed box I'd bought at a yard sale, lining it with felt and burning into the molding with a hot soldering iron the words, "This box rebuilt by J.A.C. in 1975. Given to Joe Coomer II by J. Coomer III. To keep anything worth keeping. Feb. 22, 1975." I certainly seem to have valued thorough documentation. Till the day he died, Grandpa kept the box under his bed, stuffed with his favorite things, his knives, watches, lighters. My father returned it to me after the funeral.

I brought my first friend from school, Bobby Ryerson, to my basement as a sort of introduction to me. He hit tennis balls against the concrete block with me, hit the balls I missed, and we became fast friends, allied against the wall and everyone else in the halls of Lakeland High School. He replaced Zeke in a way, and he and I in the security of our parents' homes practiced blending into the frightening walls of adolescence.

24

Our T-shirts hang limply on nails by nine o'clock, and the ice in the cooler has melted by two. It hasn't rained since Heather and I left, and in the morning when Mark and I pull up to the house we have to sit in the cab till the dust settles. We drum our thumbs waiting for motes to fall. This heat. Heather rails at the weatherman every evening and curses the high that sits directly above Veal Station, Texas. The horses come up and lean against the house in the shade. Flies lick the panes of the first windows we've put in, licking the reflected sky, water so blue and perfect it doesn't exist. It hasn't rained in weeks, I'm telling you. The tongues of dogs are obscene. Cracks in the bottom of the pond seem bottomless. Many days over one hundred degrees. Dirtdobbers come to my dripping wellhead as to Mecca. We see them taking off, flying by, balls of mud in their teeth, building houses. The sun turns my sheet-metal-smooth roof into the cellulite of a big woman's thigh. It buckles with the heat, but the nails hold. This heat.

We're plugging windows into their sockets, $11,511 worth of windows, the single most costly ingredient in our house. There are sixty-two of them, Anderson Perma-Shield white double-hung technical virtuosos of insulation, sun inhibition, frost prevention and maintenance freedom. You can also see through them. We splurged for several reasons. We figured the windows would pay for themselves in heating and cooling savings by the time we died. We'd need good windows in Texas, what with the weather's rapid fluctuations, able to last through many openings and closings. And finally, since the frames are wood inside and vinyl out, they please us aesthetically and paintingly. While Heather and I like to consider ourselves creative, we know we lack endurance and wouldn't want to paint sixty-two windows every five years. Thoreau had only two windows in his house, one on each side, and paid $2.43 for them. But his concern was economy and I have to watch out for Indians.

Most of these windows are large, three feet by six feet. Carrying them requires both of us: Mark shoves them out the open hole to me, we pop them into place, and he levels them and holds on while I hammer roofing nails through the vinyl into the waferboard. I'm fond of plugging components into place. Cutter plays with his Lego blocks, and I have my house. The first-floor windows go in quickly. It takes longer to get them out of the box than to install them. But the second and third floors require some adjustment. I'm able to stand on a long ladder to put in the second story: nail one side, climb down, move the ladder, nail the other side. The ladder won't reach the windows in the gables or the third-story of the tower, though, so I have to risk my life. Our first step is to cut arches out of the waferboard above each window. All third story windows are half-rounds above three-by-five rectangles. Then we run the window out the hole, holding on and grasping with all four hands, and finally bring it back into place. While Mark holds the window I lean out, nails falling to the ground thirty-five feet below, and tack the window into place. By raising and lowering the two sections of the double-hung rectangle, I can reach all the way up both sides but can't get to the top of the arch. So I nail a handhold on the side of the house and climb out, standing on the sill. Mark lowers the window to just above my toes and hugs my belly through the open upper sash. I start a nail, clench my two-by-four handhold and drive the nail home. After a few more nails I wheeze back inside.

Mark says, "I didn't like that."

"You're gettin' paid, aren't ya?"

"Your belly stinks."

"I'm risking my life and you say your nose is bothering you."

"It's your house," he says.

Two more hours of tendon-taut work and we finish. Heather comes out for lunch and opens and closes every window in the house. She pulls me around the house, forcing me to stand in front of every window, one by one, and stating, "Look, this is the view from this window," and "See, every room has a view. I designed it that way. In this bedroom I'm going to put a copy of *A Room With a View* on the bedside table. Our friends can visit us and sit in a chair by the window and read it."

In the afternoon Mark and I go back to the lumberyard and frame up the old stained-glass windows we've found. They're all in pretty good shape now. I managed to step savagely into the middle of one a few months back, but Phil has repaired it for me, finding glass to match at a crafter's shop in Fort Worth. We build sashes and sills and frame in the stained glass with generous doses of silicone. We're both skittish about banging around the glass with our hammers, but it seems to have some play, and we get the frames all built and back out to the house without a crack. Altering the natural light, we fill the house with shafts of color that fall upon floors and walls in symbols. I hold my hand in the beam and take upon my palm a bloody heart, a gold ring, a white cross, the head of a dog. A yellow sun isn't enough for us, not enough God there anymore, so we build figures of glass and look at the sun through that, thinking the prism father of the star behind it. A day hasn't dawned that we haven't called something else.

We put hearts up in the kitchen, a retriever in the downstairs bath, a huge sheet of cranberry glass on the third floor alongside a pair of intertwined gold rings given to us at our wedding by an old friend, and in the front stairwell we install a door that no one will ever walk through. It's one of a set of three art deco doors that are four-fifths stained glass. They're English, perhaps from a church because there's the outline of a cross in the glass, but mostly they're deco: in line, color, geometry, they scream 1935. We know they aren't the correct period for our house, but their beauty overcomes time. Two of the doors will open off our master bedroom verandah, but this third one will hang suspended in our stairwell, thronging the well with colored light, a small chapel to rise in.

We square up all the old exterior doors next, removing some of the influence of their original houses, bringing them into line with our house. We'll at least start square. Someday, as my house sighs and swags, I'll pare them untrue as the door owners before me did. And when someone salvages the doors from this desolate structure, if there's anything left of them, they'll square them up again. We strive to level our dwellings, but so does Nature in her way, and there's the beauty struggle, keeping our bubbles centered, the ocean with no breeze. We hang all the exterior doors: the cows in the back door, the Persian scene double front doors and the stained-glass double doors in the master bedroom. I say this in a sentence but it takes all day. You can hang a man in half the time it takes to hang a door, and do it with much less cursing. A man hangs from a single rope, plumb to the world; a door hangs from two or three hinges at any angle it wants to. It can scrape the floor, scrape the wall, close two feet from the room it's supposed to shut off, and open of its own accord. It is a thing alive long after it's dead. Mark and I battle with the level, the hinges, using successively bigger nails to fasten the frame of the door to the house, and finally are able to force the doors into some agreement. When they're all shut we've closed in a piece of the world. Wind can only enter through cracks where sheets of waferboard meet. The house whistles through these cracks, and our doors, caught in a vacuum or still full of old ghosts, open. And later, when we're engaged with something else, they slam to, ripping our breath from our lungs.

Through August we work on the ornament of our house, the Queen Anne of our Victorian. It's all outside work, and our first priority is to keep the rain and wind out. We begin with porches.

Beyond their help in cooling the first floor, they define beyond a doubt that we're building an American house. When I sold my first novel, I took the advance and traveled for three months across Europe. Not once did I see anything resembling a front porch. I don't know if Americans are nosy and like to watch who's driving by, or if, even after two hundred years as a nation, we're still uncomfortable indoors, preferring the halfway house of a roof and posts. Porches have always seemed to me to be places without rules, where anything might happen. When you step in someone's front door, there's a prescribed list of cultural mores, but a porch has none of this, almost as if it were outside of society. The opposite is true, of course: the porch is usually a hotbed of social activity. And not just human society. Dog and bird are as comfortable and welcome there as homo sapiens. I find porches places of great joy. We shout at people from our porches, wave at them and invite them up for a swing and a glass of lemonade. It's a wonder more children aren't conceived on porches. I was almost killed on one, but the nature of love for porches in my family is such that we've turned the incident into some kind of comical myth and legacy. When I was only several months old, my Aunt Melody dropped me on my grandmother's concrete front porch on Refugio Street. I landed on my noggin, still soft but it rang like a rifle shot. I cried a lot, I'm told, but bled only a little, so they didn't take me to the doctor. A month later when the cut on my forehead had healed, my mother noticed that the scar was still blue. A small piece of coal had embedded there, the skin had healed over it, and in a million years when I'm archeologically excavated they'll find a diamond on my forehead and think me a minor king of my time.

The back porch is very simple, a shed roof covering six feet of concrete. We attach the rafters to the second-floor joists and support the far end of them with four posts. The roof has a shallow three/twelve pitch, easy to walk on, high enough to let the late evening sun enter the stained-glass windows of the kitchen. We nail waferboard to the rafters, box in the eaves with one-by-eight redwood, slap on some one-by-two tin edging and roof the porch with tar paper and red composition shingles. "Composition" is a euphemism for "tar and rocks," but the shingles shed water under either name. I set my father's radial arm and table saw under the porch.

They'll remain here for, I imagine, some time, out of the house but out of the weather.

The front porch is on a completely different end of the house. Facing the pond and the east, it will wrap around the tower and continue halfway down the south side of the house to the dining room. This will give us plenty of room for swings and rockers and should provide at any time of day some point of countryside perfection, a precise combination of breeze, sunlight and shade. I'm talking about lying down on the swing and being able to fall asleep immediately. We have to put this roof on a four/twelve pitch to make it look right. The difference between a four/twelve and a three/twelve is that hammers slide off one and not the other. I have never found this information in any carpentry manual. The east and south sides of the porch are much like the back porch, a simple sloping shed. It's the bit circling the octagon tower that's geometrically difficult for our minds and saws. To butt the rafters smoothly to the ridges coming off the corners of the tower, we must cut impossibly sharp angles. Our power saws don't have deep enough blades to manage it. So we get out the old handsaw. To date we've only used it for simple cuts in hard-to-reach places. It's still almost brand new and very sharp. The Victorians built entire houses using one of these things. We mark our line on a two-by-six and I start a cut the way my grandfather taught me, on the backstroke with my left thumb as a friction guide. I hold the saw up in the air and smile at Mark. I waggle it to make thunder. Then I begin. After eighty-five or ninety strokes Mark says, "Well, let me try it." After he wears out I spell him, and after he spells me one more time we finish. Only seventeen more of these to go. There is nothing so humiliating as to begin a saw stroke, have the saw bind, and then to try the muscle of your arm only to have the saw U-turn, throwing your whole body up into the air. This happens to us several times on this one board. We marvel at the strength, patience and skill of Victorian carpenters. The rafter butts up smoothly to its ridge, and we nail it to house and home. We cut the remaining seventeen rafters as sharply as possible with the circular saw, using extra nails to hold them to the ridge. This leaves huge gaps in the joints, if you can call them joints, but our arms are thankful for it, and who will ever know, if no one tells, once we get waferboard on top and ceiling beading below? I doubt if I'll even remember it in ten years. We waferboard the porch then, cutting huge trapezoids for the sections around the tower. As an afterthought of design we add a small gable to the porch, above the front doors and the same width as them. Then the trim and metal edging and tar paper and the drudgery of carrying bundles of shingles upstairs and out to the porch roof.

After that and a Big Red it's fairly pleasant work, tacking down the shingles, careful of the lap, catching my hammer before it slides off the roof. It's mercilessly hot even though the trees that virtually surround the porch shade us. I settle into my work, the rote of it, lap, tack, tack, tack, watch the tack not your thumb, lap and realize that Don Quixote and Charles Darwin are jumping, together, from their boxcar before it comes to a full stop. They hit the ground hard and roll down the embankment, Quixote cursing, Darwin unbelieving, toward the dry ditch. They tumble over the grassy edge, breaking off a piece of the earth's crust, and fall into the red world with a slot of blue. They fall for two and a half feet, straight down, falling for centuries. The dust settles slowly over their embrace.

Quixote peeks through the canyon of his armpit and shouts, "Arrived!"

And Darwin rolls Quixote and his armor off his legs. "Ape," he mumbles.

"Up, up," Quixote orders.

"We've got all the time in the world," Darwin ruffles, brushing the seat of his trousers, smoothing back his hair.

"No," Quixote counters. "Things change."

I'm going to have to build some bleachers, I think. They may get here early. So while I roof I construct a set of bleachers out of two-by-twelves and angle iron and put them on the south side of the house under two oaks. I build bleachers, but by the time I'm done they resemble pews. There, I think. I'll have to get hot dogs too.

Our last job before we can begin siding the house is the most miserable: boxing in the ends of the rafters and installing air vents. Every wasp in America is waiting for us to get this done. They'll all build their homes there, where you can't reach them even with a long stick or the water hose. The work is miserable because you have to do it all above your head, while you stand on the end of a thirty-two-foot ladder. There's no one else up there with you to hold the other end of the board while you try to get a nail started and driven. So you use your head as a third hand. I get two pieces of soffet up like this before I'm besieged with a headache and a knot on my skull. A brainstorm blusters in from the trenches of France. Back at the antique mall I buy Mark and myself two World War I steel helmets, and we assault my house under their protection. We wear them for two days putting up the soffit and soffit vents. I become rather attached to mine, having missed action in all wars, and wear it when I eat my K-rations and when I drive my tank home. I think they should be standard issue to all sheetrockers. Besides protecting your head they give a certain sense of pride in work. I am almost as selfish with my helmet as with my hammer. I feel I have fought the good house.

Our siding slides off the delivery truck in a long, grey, limp bundle. We've chosen three-lap Masonite siding. It makes me want to aesthetically puke, but real wood siding is cost-prohibitive and probably wouldn't last as long as the Masonite, which is fairly impervious to rot, dry or wet. The Masonite is also primed a battleship grey, which means we'll only have to give our house one coat of paint. This is an immense savings in time and paint, considering the size of the house and the professional painters we'll hire: me, Mark and Heather. The siding is a foot wide and sixteen feet long, and the three laps simulate Victorian four-inch siding. It's fairly convincing from a distance of thirty feet and beyond, in the way that a boat looks like a duck. But it does shed water almost as well as a duck, and that's what's important now. This siding goes up from the bottom the same way shingles do. The only step that requires any real effort and conscience is the first row around the bottom of the house. It has to be level or the whole house will either slide down the hill into the pond or look as if it's about to become airborne. Though this last notion appeals to me for a moment, Mark and I work our way around the house with the siding and a level. This takes a couple of hours, but what follows is an unprecedented degree of psychological accomplishment. The siding is smooth, ruled, ordered, and nails up quickly. It covers the motley shapes and colors of the waferboard that we've been looking at and working with for months. The house begins to look like a house instead of a great brown barn. The siding provides us with so much satisfaction that we work through lunch, as though we could finish the whole house by evening. But it takes several days. The higher we go the slower we go, climbing ladders, dropping hammers. By the time we reach the eaves we're sick of Masonite, even the smell of it. Our great brown barn is now a great grey battleship on the plains. We understand why the Victorians used so much ornament on their big houses: simply to break up the vast monotony of horizontal line. Our house could be a piece of sky through Venetian blinds. If it didn't have gables and bay windows it wouldn't look much different from the Great Wall.

Above the level of the eaves, in the gables, we begin with our ornament. After erecting twenty-six feet of borrowed steel scaffolding under the north and south gables, and then borrowing my father's air compressor and staple gun, Mark and I throw up row after row of cedar fishscale shingles. These go up quickly as well, especially with the staple gun. The only thing that slows us is the way the scaffolding drunkenly weaves, and our memory of the eight or ten short pieces of lumber we had to stack under the scaffolding legs to level them. Mark finally ties the scaffolding to the house with ropes. Near the apex of the gables we cut a hole for a round redwood vent, tack up the last few shingles, install the vent, climb in the house through a window and race downstairs and outside to look upon our work. I can't wait to get home and tell Heather.

We'll try to make the east and west gables—the front and rear of the house—our gingerbread showcases. We haven't found any gable gingerbread on our salvage searches—it would probably be severely weathered and the wrong roof pitch anyway—so we'll make most of ours from scratch. We begin, once again, with scaffolding. Half of the time we work is spent only in preparing to work. Even with this we'll have to stand a ladder on the scaffolding to reach the peak of the roof.

In the evenings, Heather and I pore over our snapshots again and go through all our Victorian books and magazines, looking to steal ideas on gable decoration. Sunbursts seem to be dictated, as well as decorative shingles, and finally some piece of gingerbread hanging from the vergeboards that we'll just have to come up with from what parts and scraps we have and can invent. An arched window sits in the middle of all this.

We start with a row two feet high of lap-and-gap siding laid vertically. On top of this, radiating from the bottom corner of each side of the window, we build quarter-circle sunbursts, laying rays of Masonite over one another so our sun will shed rain. Over the remaining surface of the gable we alternate rows of fishscale and arrow shingles to form rows of circles. Another round redwood vent completes the waterproofing of this end of the house.

A botched piece of fretwork on a Victorian house can contribute a great deal to its overall stupidity, so I worry a great deal. I draw the outline of the gable and window on the concrete floor of the living room and with my lumber pencil sketch several designs on scrap lumber, trying to utilize some of the old furniture parts from the Pittsfield auction and a box of short, turned porch pieces from an auction in Fort Worth. Heather likes the outline of a decorative piece off a dresser that typologically could be a snake, fish, whale or sperm. Since we're thinking about calling our house Rainwater, I cut a big raindrop pattern out of plywood. There seem to be enough turned pieces to follow in a half-circle the contour of the top of the window. And from the Pittsfield box we pull out two pair of turned walnut finials, probably removed from an old bed. We cut several rain drops and spermatozoa from two eight-foot two-by-twelves we've set into our gable pattern on the floor, then attach another, shorter set of two-by-twelves to the first pair, and screw the turned porch pillars to a semicircle cut out of them, bringing them together at their bases in a fan. Then we glue, screw and nail the whole mess solidly together and let it dry overnight. It's all very exciting, the most creative work we've done yet, and if I could get steady work in this field I'd give up all my dogs.

In the morning Heather comes out with her camera to record the ascension of the gingerbread, the flight of the phoenix. The conglomeration is too big to go out the third-story window. Mark and I carry it to the second floor and out the double doors to the porch roof. We find that gingerbread is not only awkward but heavy. We tie ropes through two raindrops, and while Mark holds it steady, saying over and over again, "We'll never be able to do this, there's no way we can do this," I run upstairs, step out the window and have him throw me the ends of the ropes. Mark comes up then and we try to lift the fretwork to our level. Heather stands in the front yard, camera poised.

"Lift!" I yell.

"Lift!" Mark yells.

We lift, and with a great many "I can't believe it's this heavy" exasperations drag it far enough to get a hand apiece on it and pull it up to sit on the scaffolding. The first thing Mark says, when he's able to catch his breath and has wiped all the sweat off his face, is, "We'll never get it into the gable." We discuss several options, finally deciding that we have none, must raise it and screw it to the house with brute strength. "Who's going to hold it up while you're screwing it to the house?" he asks.

"I'll screw real fast," I say.

"I could be digging ditches right now," he says. "I could be going to grad school."

We gather long screws, the electric drill, ladders, hammer, nails. Heather watches us with an intense curiosity, the camera limp at her side. I start four screws in the fretwork, Mark and I hitch up our pants and take deep breaths.

"You've trained your whole life for this moment," I tell Mark. I put the drill in my nail apron, and together we carry the fretwork up our ladders, balanced on the scaffolding thirty-two feet above the United States of America.

"Okay?" I yell, the fretwork in my hair, teeth, asshole.

"Okay!" Mark curses.

I let go with one hand, holding the gingerbread up with the other, and with the strain my body is under it takes me several seconds to get the screwdriver bit to the head of the screw. I keep missing, breathing hard. Mark can watch me. Beads of perspiration course down his forehead, and he's blinking a lot. I miss yet again, the drill glancing off the screw and ramming the fretwork.

"Goddamn uncoordinated son of a bitch," he screams.

Finally I get the screw halfway in and can relax a bit to drive in the second screw on my side. Then I climb off my ladder and up Mark's, leaning against him to screw the other screws in. We come down then, holding our hands above our heads and crouching, afraid that the gingerbread will fall on us now that we're completely exhausted, until we see that Heather is taking our picture. We stand up straight, face forward. When we hear the click of her camera we return immediately to our cowering positions and put thirty-five or forty more screws and nails into the gingerbread and gable. When it's secure we drill holes and suspend the finials from the lowest points of the fretwork. This makes the gingerbread seem as if it's melting, dripping off the house. It's beautiful. The sun shines through the open work of raindrop, spermatozoon and turned work, and shadows are made against our house, and in our house through the window, and even though I know Thoreau might not be pleased with all this adornment, I cannot help but think he would have liked the interplay of light, shadow, shape, and pleasure in our eyes.

On a whim, at the end of the day, Mark and I cut a crescent moon and several twinkling stars out of sheet Masonite and nail them under the gable window. It's a decorative effort straight out of the Middle Ages and the 1960s, but when we come down and look up at it, the moon and stars seem at home with the sun and rain, and we have made a heaven of our house.

The gable at the rear of the house follows in the same manner as the front: scaffolding, vertical siding, sunbursts and shingles, but we like the creative effort required of new gingerbread: instead of raindrops and spermatozoa we through-cut teardrops and cattle. The cows walk down the fretwork to the initial H on one side and J on the other. I drilled the initials with a half-inch paddle bit, but I will tell people I shot them out with my forty-five.

Over the next week we shingle the tower, and cut small fretwork pieces for the eight corners of the tower eave, connecting them with small turned spindles. Then we nail up all around the house a one-by-twelve decorative molding band separating the first and second floors. It helps break up the large areas of siding. Our final job before painting and applying trim to the windows and doors and corners of the house is to shingle the four small dormers sticking out of the roof. I decide to let this wait till we've picked out and tested our paint scheme, since I'll have to move the big roof ladder again just to paint. If I side and trim and paint all at once, I'll risk my life considerably fewer times. And besides, the waferboard will keep most of the rain out till we get there.

I am almost through with worrying over water, bless me. And it's a good thing: even though it's early September and ninety-five degrees out, Texas is nigh onto winter. It will happen one evening between eight-thirty and nine P.M.: a north wind and a degree drop unthinkable, three shirts and a sweater the next morning, hanging up plastic with bitter hands. I can't wait for the day when I take hammer in hand and smack my half-frozen thumb.

In the middle of the month Mark and I lay Italian tile on the front and rear verandahs, sloping it to the porches. The grout is red, as the tile, and our hands are pink for days afterward.

On September 19, 1987, we spend our first night in our house. Mark and Tina join us. Even though the house is ninety-eight percent watertight and there is no rain forecast, Heather and Tina insist on putting up two pup tents in the master bedroom. Mark and I hammer sixteen-penny nails into the floor for tent stakes. We eat meat and potatoes cooked in foil over the coals of scrap lumber and watch a silent movie (Marion Davies in *Showpeople*) on a bug-caked twelve-inch black-and-white TV. Later as we lie on our sleeping bags with the mosquito nets zipped up we laugh at the unspeakable talents of the Miss America Pageant. When the set is off, we listen for the night sounds, coyote, whippoorwill, cackle of bug and croak of frog. I try to notice some significance, to feel some moment in its passing, our first night in our home, but the most I can come up with is Heather's giggle in the pup tent, the warmth of her breath and smoothness of her body. In the night I realize that we are together in our house, that we build to ensure that we hear each other's laugh clearly.

The morning begins so early I hardly recognize it.

25

We had an emotional attachment to Kentucky long before we moved there. My father and grandfather spoke of it as the promised land; faith gripped their voices. Even though they left

Breathitt County in 1947, when my father was only eleven, they still considered it their home. Phillip and Sally and I were simply glad to have our dog back. Zeke, who'd been shipped from New Jersey to Indiana to Texas to live with my grandparents, finally came back home to us. My father bought us seventy acres of farm, creek and pond to play on. It cost him $72,000 and two hours of commuting time to Lexington every day for the next four years. IBM had transfered him again.

I never knew that a home could be such a big place, that it could include, for instance, a creek, or a cow. But it could, and did, as much as if the creek ran down the hallway and the cow ate out of the sink. The variety and originality of our lives' moments increased and built upon one another to the point that we ceased to recognize them, simply lived in a perpetual state of awe and dumbfoundedness. Our seventy acres lay literally across the street from the Appalachians. On a state geological map Highway 34, Lebanon Road, marked the western boundary of the mountains. Quirk's Run (we named our farm after the creek which formed most of its boundary) was a series of three grassy hills nestled into a bend of the creek. There were a few trees, mostly along fence lines and in the creek bottom, but for the most part it was lush grasses on steep hillsides. Fifteen acres along the state highway formed the nucleus of plowable land. We raised clover hay on most of it, tobacco on four acres and a garden next to the house.

We were five miles from Danville, a town of about 10,000 then, and forty-five miles from Lexington, 50,000, but only a couple thousand feet from Parksville, population perhaps thirty-five or forty. The railroad passed through Parksville, quickly, and Feather's General Store was there, a one-stop for almost everything we ever needed, even though the store was only twenty feet wide and fifty long.

The basic character of our lives changed. No longer did we load up on weekends and visit every sight within two hundred miles. Rarely did we sit in the house on a Sunday afternoon watching a Dallas Cowboys game. We began by exploring. For me this exploration lasted for the whole four years we lived there, and continues to this day. I still think about that place often, trying to figure out what it did to me. Perhaps it was only that I lived there from the ages of sixteen to twenty and these are impressionable years. But I think it was more than that. I think all my past and all my future caught up with my conscience there, and I felt lucky for it. I felt I was in the right place at the right time, somewhat rare for a teenager.

We walked the creek first, our whole family, in a primeval, territorial way, along its bank, through brush and around great hickories and walnuts and sycamores, their roots exposed and hanging into the water. An old post road followed the creek, although trees with trunks six inches across grew in the middle of it now, and its wooden bridges, crossing and recrossing the Quirk's Run, were long since rotted away. The water ran clear and quick over slate and polished stone and dropped into bends where it slowed and deepened, became green and murky. We stepped over turtles, startled rabbits. I remember putting both my hands in the cold, running water, lifting a flat rock, looking for crawdads. My father tugged at barbed wire buried inches deep in the trunk of a tree. We found the skull of a cow. And I realized, in the cool and shade of the massive trees above us, realized somewhat peripherally, that my past was older than I'd consciously thought.

Every building on our farm, and there were many, stood in the four-acre southwest corner near the state highway. There was a massive tobacco-drying barn, a small stripping shed, a ten-stall horse barn with hay loft and tack room, a metal five-car shop and garage, an outhouse, a chicken shed, a log cabin built during the twenties as a child's playhouse, several other derelict shacks, a cellar and finally a large neo-Colonial house built by a Dr. Brown in or about 1906. The house was known locally as the old Dunsmore place.

Although the house was livable, it was only barely that, and we immediately began to rebuild it by tearing great chunks down. The original house had been a Colonial front with four rooms, two above and two below with a hall and stairway to separate them. Each room was perfectly square, twenty-five by twenty-five, and had a fireplace. The two chimneys rose up on either end of the house. A brick front porch one story high ran the entire length of the house. Later someone added two additional single-story rooms to the rear of the house, an ell. Porches were built onto both sides of the ell and a fireplace onto the rear. Still later, perhaps in the forties, the porches were closed in and one was turned into a "modern" kitchen and bath. These rooms were all of five feet wide. The entire house was covered in a red composition siding resembling brick. All five fireplaces were closed off in favor of kerosene heaters. Still later a fifteen-by-fifteen bedroom or pantry was added off one of the closed porches. All of the additions were, I surmise, done with alacrity and a lack of skill. Windows still sat in interior walls, separating one inside world from another; no moth had fluttered against them in thirty years. The bathroom floor was caving in, as well as the ceilings in the pantry and kitchen. A new roof was required over the additions, and all of the wiring had to be replaced.

With crowbars and hammers we completely toppled the pantry. We opened both porches back up again, so that for several weeks my mother cooked all our meals out of doors but on modern appliances. The bathroom was also open, save for its porch roof and bedspread walls, for almost two months. In the meantime Dad hired a bulldozer to fill in the concrete and rock cellar. Over the fill we built our own addition—two bedrooms, a bath and laundry—and used the old back porch for a hallway. This was my first experience with house construction. Dad would do the greatest part of the work on weekends, but during the week he'd give Phil and me jobs to work on. We dug the foundation trench, nailed down the plywood floor and with Grandpa Dennis's help shingled the new roof. A group of my father's friends came down from Lexington on a Saturday and worked all day sheetrocking the interior, installing a full beer can in a wall, a note with all their names attached. Phil and I spackled the ceilings of our own bedrooms, in a random sponge pattern that I'd like to go back to see now. I picked out my own wallpaper, one of my last acts of childhood, choosing a pattern of old cars and motorcycles. We covered with white aluminum all the new work and the old brick siding, excluding the real brick front porch. Our house joined hundreds on Lebanon Road, white with green trim, a black tobacco barn beyond.

Dad had electric baseboard heating installed, so we sold the monster kerosene heaters and opened up the living room fireplace again. We paneled or papered and carpeted every other room in the house. Mom got a state-of-the-art kitchen in the older addition: new cabinets, appliances and tile flooring.

While we worked we explored, shining flashlights into recesses and walls, and found old newspapers, a few bottles and under the staircase a whole tray of arrowheads and stone hatchets. Indians!

When we finished with the house we began on the yard. It was completely overgrown. The previous owner, a divorced farmer, had simply mowed a fifteen-foot path from the back door to a huge trash mound behind the house. We burned several rotting shacks where they stood, carried and dragged the rest, load after load of debris, to a ditch on the back of the farm and burned it there. Later we built a board-horse fence down the quarter mile of highway frontage and painted it white. It looked swell against the green grass.

We attempted to make the farm pay for itself, trying to squeeze money out of dirt, and most of our efforts were successful to the degree of making us miserable. My father bought calves. It seemed a simple thing: to let them eat grass. But as it turned out they required inoculations, vitamins, salt blocks, hay in the winter, and as some of the calves grew older we had to deprive them of their bullhood and remove their testicles. I'd never, in my suburban life, considered cows as curious animals, but they like to see the world, roam beyond their fences, especially in the early morning hours. My father and Phil and I chased cows continually it seemed, rising from bed with a phone call from one of our neighbors, cursing any sacredness a cow might possess.

My parents kept a garden through the summer, a lackadaisical garden at best, mostly of weeds but with an occasional tomato or cucumber. These by-products of the earth we looked upon in a kind of stupefied amazement. We held cantaloupes in our hands as if we'd never seen them before. It seemed strange that the planet could focus enough of its energy to design and build an ear of corn out of water and dirt. We began to realize that we'd been out of touch. We grew an acre of feed corn, letting it dry in the field till it screamed with the wind and became an acre of knives, then tromped through it with heavy coats and gloves, breaking off the ears and throwing them in the back of the truck. I always liked this work. It seemed like classical harvesting.

I walked the Quirk's Run and the fields and pastures above it hundreds of times over the next few years, sometimes alone but almost always with Zeke, who felt the same responsibility I did, and often also with Phil or my father or a friend from high school or college. I walked it with friends in a state of pride, showing them owl nests and perch in pools and the stone bulwarks of the old bridges. I had most of my confidence there and felt persuasive beyond my abilities. The farm became, for me, another member of my family. On weekends from college I returned to it as much as to anyone else. I know that piece of land better than I do any other on the planet, including the farm we build on now and the fields around Saginaw. I miss it in the same way I miss people, and have irrational urges to get up from my chair and drive to Kentucky just to be near it. I have joined the faith of my father and his father.

26

We're playing a game of hide-and-seek, but the sun is always It. Through the day the slow and still persistent September sun chases us around the house as we paint. In the shade it's already fall. Heather has a blue, long-sleeved shirt on over a white T-shirt, and her hair is bound in a pony tail by a rubber band. I have to stop myself from climbing down off my ladder and chasing her. She paints as high as she can while Mark and I stand on ladders of varying heights, brushing as far as our arms will reach, buckets hanging from homemade S-hooks. The house is coloring by shade and shades from the battleship grey of the Masonite to the muted grey/beige of our mind's eye and Mooreguard paint. We can't use rollers because of the ridges in the siding and can't use a spraying rig because of the many windows, and so we paint a house the way they did a hundred years ago, with brush in hand. It's slow, peaceful work, fine work for a shady fall. It's satisfying too, for the accomplishment-oriented. You can paint a wall in almost no time, and the results are complete and immediately recognizable. Painting seems to calm the frenetic banter of construction. The work is so simple it's numbed our speech. Once I've trained my eye to tell my hand to make everything grey/beige my lips only move when I breathe.

I'm thinking that the change is barely noticeable but that it's sure: the season's turning. If it were summer we'd still be burning, even in the shade. But I haven't begun to sweat. Inexorably I force myself into that by now old and tired realization that time is passing, time has passed. This always happens to me in the fall for some reason, at the first hint of coolness. Perhaps it's just an old instinctual shudder to find a cave or store up a few extra acorns. Perhaps it's because my birthday is in the fall and this year I'm approaching thirty. My parents say that I was old the day I was born. But I'll only be twenty-nine this time around. What's thirty to me?

I've never painted a house before. Almost all the houses I've lived in were brick or stucco or aluminum. Thoreau never painted his cabin. He sheathed the studs with feather-edged lumber and then shingled the entire house, roof and four walls. Shingles bear the weather well and show their age with a grey distinction. If Heather and I hadn't built a Queen Anne, we probably would have chosen a saltbox and used the same siding Thoreau did. The most romantic and beautiful and intriguing buildings I've seen are derelict houses and barns, bereft of all paint or stain, weathered by wind, rain and sun, the lumber showing its grain three-dimensionally. Your hand can see the years as well as your eye can. Weathered lumber has a past, and, strangely, some boards in their erosion look almost human; they almost look weary. My hand constantly reaches out to touch them, to know that age. But Thoreau lived in his house for only a couple of years, and it lasted as a house for perhaps twenty-three years before it was dismantled and used as patching lumber. I plan to live in this house for a good many years more than that. The Masonite is a great help in this endeavor, since it has a half-life of a million years or so, but an extra coat of paint couldn't hurt.

Queen Annes require paint for aesthetic reasons also. Paint is used, in several complementary colors, to pick out the architectural detail. There is nothing sadder than a Queen Anne dressed all in white. You can't tell her eyes from her knees. We'll use three colors in addition to the white of our Anderson windows: grey/beige for the body of the house, a brick red for the eaves and the belt around the house and a dark grey for the door, corner and window trim. We'll use all three in the gables: red for fretwork, dark grey for the decorative shingles, and grey/beige for our sunbursts. These are military colors, I suppose; perhaps we should hang a medal or two on the gable. We arrived at these three through an experimental process, using the front gable of the house as a work table. The shingles there have four coats of paint already, and if some unlucky historic preservationist of the future should do their detective paint work there, they'll be completely baffled. But they might get a good laugh out of it. Our first effort resembled the American flag. Our gable was red, white and blue, and although I feel a sense of patriotism, these colors made our house look like a parade float. In our next two attempts we fell victim to the temptation to pick out too much detail with our paint: in the first the house looked like some kind of sixties hippie kaleidoscope; in the other the gable was a World War II Japanese battle cruiser complete with rising-sun flag. We finally came to the muted colors we paint with now, letting the architecture speak for itself.

While we paint, the sun works its way around the southern exposure of the house, so we gather our ladders and buckets and move in a northeasterly direction, holding our brushes to the house, seeking more shade. When we finish a large enough block Mark and I lay out long one-by-four and one-by-six redwood boards on the sawhorses and give them two coats of dark grey. As Heather continues around the house, painting as high as a five-foot-tall person can paint, we cut and nail up this simple trim around the doors, windows and corners. Heather has to stop painting to walk around the house and inspect every board we put up. I don't know if it's to relieve the monotony of beige or because each piece of trim completes that portion of the house and she wants to see how our notion has come to pass.

This expanse of beige. We've emptied bucket after bucket, worn out half a dozen brushes. The earth around the well spigot is grey/beige too from our washing out brushes at lunch and end of day. We each have two sets of clothing that are now grey/beige, not to mention the semi-permanent grey/beige stains beneath our fingernails and in the recesses of our eyeglasses. I try to remember how Tom Sawyer got out of this, but his method won't work here; everybody else is already painting. If I pretend it's fun, they might leave me to do it all in my happiness. So I just paint. Left, right, left, right, and every once in a while I'll come to a recessed nail head and I can pound the bristles into it. These are days of repetition, knowing each evening that there's nothing but more paint for the morrow.

But on the morrow things become interesting. We arrive to find a pigeon murder in the master bedroom. Feathers are everywhere, and they scatter as we swing open the stained glass doors, revealing a horror story on the floor: a spot of blood, a spot of feces an inch or two away, and all else save feathers gone. Whether the kill of owl, cat or coyote, we don't know. We still leave doors and windows open; there's nothing to steal or break. Immediately Heather thinks of the two squab being raised in the attic, but we find them safely in the nest. The victim must have been one of the four or five adults always poking around the porches and ridges. The pigeons' favorite perch is the very peak of the tower, astride our upturned Indian vase. They remind me of seagulls on piers and leave the same familiar streaks on the tower copper. Another couple has built a nest on the front porch roof, a sad choice because their eggs roll off the sloped roof and break as soon as they're laid. We listen to the cooing and gurgling of these birds, orgasm after orgasm, as we paint. The birds walk along the metal roof, clickity, clickity, and peer over the edge at us, turning their heads to the side the way a dog does when intrigued. Heather screams at them because she hates the mess they make on our newly painted house, long trails of green and white ending in a dark brown bugger that hardens to stone if you leave it too long. But we're letting the squabs come of age before we begin locking the house up at night.

Earlier in the week we had a meteorite. We didn't see it, but the TV weatherman tracked it on his color radar for three blips till it either got too low or disintegrated. It was tracked over Denton County, heading southwest, so maybe it fell in the field below our house. But we've found no smoking fragments or metals not of this world, only a slightly dazed cat that stalks Blossom and the dogs when we take them out to the farm. Perhaps the meteorite struck me, or maybe it's just the fumes off this paint.

I look up, out across the field of summer dry grass, and see Emily and Thoreau there. Thoreau steps along slightly stooped, his hands clasped behind his back, looking down at the ground.

Emily sighs. But sighs, and looks up at Thoreau. "Our Saviour after the cross, after his knowledge of failure," she says.

Thoreau nods.

They walk on, silently, till Emily notices Henry's demeanor, the wingbeat of birds in his eyes.

"I never got out of the house much," she softly says.

Thoreau looks up too. "I always said one never had to."

"Me too," she says, "but . . . this place. We have some time don't we?"

"Of course," Thoreau says."We're early. I mean, whatever we decide to do, we'll always be on time."

"Well," and Emily uncoils in the blue sky, twirling in her cotton dress, her cheeks burning. She spreads her arms, turning, and says, "To make a prairie it takes a clover and one bee, One clover, and a bee, And revery." Turning, and Henry David, thirteen years her senior, stock-still, in love, watching her turn. She spins and spins, reciting, "To make a prairie it takes a clover and one bee, One clover, and a bee. And revery." And then she stops, nearly falling, a caught giggle. She looks around, turns around looking, slowly, and turns back to Henry. She holds out her open, timid hands. "The revery alone will do, If bees are few."

It must be a full moon. Time is passing. I paint faster, but that doesn't seem to slow time. I look down at Heather as if she'll never be there again. We finish with the beige and move on to the other three large gables, jamming the bristles of our brushes into the cracks and corners of the shingles, ramming dark grey down their throats. As we finish each gable the scaffolding comes down and our house has become something other than what it was. It becomes form and line, slash of unnatural color among green trees.

Soon I will have to change my way of looking at the world, from outside looking in to inside looking out, from a view of slatted, vertical darkness to an expanse of land and light. It will be strange to work inside, to not have to search for a handhold, to give up scotching one leg of my ladder, to labor without the glare of the sun in one eye and a gnat in the other. Caulk won't be able to cover my mistakes but I won't have to worry whether or not my work will shed water or stand up to the sun. I will have to drink a Big Red over this change of mindset. I might even have to sit down. I might even have to lie down and go to sleep and wake up a new man, able to put my two hands on a window casement and look out without jumping.

Time is passing. There's nothing newer than the coolness of shade upon my skin. The change has struck Mark as well. He never says anything, but Tina tells Heather over the phone one evening that they're moving to Utah in two weeks. "Hasn't Mark told you?" she asks. "We've known for a while."

"No," Heather says. "He never said anything to Joe."

"What?" I ask, when Heather gets off the phone.

"Mark and Tina are moving to Utah in two weeks."

This surprises me and does not surprise me. I am as close to a best friend Mark has, I suppose, but he was married to Tina for two weeks before I knew they were even engaged. The boy has isolationist tendencies. He's more like Thoreau than I'll ever be.

"Why Utah?" I ask.

"Tina doesn't know," Heather says.

"Does he have a job there?"

"No. He's going to look for one," she says. "Does he have any family there?"

"No," I say, and continue to eat my cereal. "It's an adventure."

"Yeah, well," Heather says. "He's got Tina and the boys now, though."

"Women and children can be in adventures," I say.

"Who's going to help you work on the house?" she says.

"I'll work on it by myself. We're running out of money anyway. I'll get Mark to help me with some heavy things and with things that need two people before he goes, and then I'll do it by myself."

"I'll still be helping," she says.

I sit there in the living room of our garage apartment listening to the cereal crunch in my mouth. Then it occurs to me to say, "Utah?"

And Heather says, "Yeah, Utah."

In the morning I walk almost as a child into our new shell. We could live here now. Our house is as firm and livable as any house was during the first years of Veal Station. We have, at last, some refuge from the Indians. I could now store up clods in the corner against some future attack.

I have determined to make the best use of Mark in what little time I have left of him. He will regret his loose-lipped wife. "Utah?" I say when he arrives. "Where at in Utah?"

We can both look through the walls. There's little room for him to hide. I'm not asking where, really, but why. Why aren't you me? I'm baffled by anyone who'd do something I wouldn't do. "Where at in Utah?" I ask, asking, Why dontcha want to build a house down the block from mine and be my buddy? C'mon, let's play catch. Let's unite against our wives. Let's dangle our dicks from the third story and talk about how cold and deep the water is in the pond below. Let's don't go to Utah. Let's not let the past become the past.

"Provo," he says.

"Why Provo?"

"They've got a good school there. I'll go to grad school. It's the mountains. I like the mountains," he says. "They have very little government intervention there."

I listen to this, trying to make sense of it. The school part is ridiculous, and Mark's favorite place on earth is Key West, Florida. It's the government intervention part that convinces me he's going.

"Have you got a job out there?" I ask, taking the father's role.

"Nope."

I want to grab him by the throat and throttle him to a state of worry.

"Just going to get up there and look," I say, explaining his plans to him. I want him to elaborate, but he just nods at me as if I've finally gotten it. He'll go along with anything as long as he doesn't have to tell me it's none of my business.

"Okay," I reluctantly add, as if it all required my approval.

But Mark is already deep into our next subject. At my father's hardware store we picked up ten boxes of twelve-two house wire and a little stapled book entitled *Wiring Your House.* The book contains wiring diagrams for the inept. The first three pages, in large block letters, state various messages that all focus on the warning to turn the electricity off, including instructions on what to do with a human in lump form, one who's disregarded the first ten pages and been shocked. We skip most of this and jump to the diagrams, looking for enlightening phrases like "black wire to green wire." This isn't a book of electrical theory or even law but of color matching. The colorblind would find it useless.

We've had electricity almost since we began, running extension cords out to the utility pole, but it's time to go between the walls. There's no danger of shock until we hook the house to the pole. So now we sew the studs together, using an electric drill and paddle bit for a needle and copper wire for thread. Everything leads back to a two-hundred-amp breaker box in the pantry that I'll let Dad, who's had some experience, wire up. The new bit cuts easily through the spruce, leaving a clean three-quarter-inch hole. I leave mounds of curled shavings that Mark drifts through, dragging wire through the new passages I've opened. The twelve-two wire coils up out of its tight roll and threads through the first few studs easily, but the more holes I drill the more Mark has to fight. When I notch out a corner and start down another wall, he can only barely get the wire through the corner cut and sits on the floor tugging on the exposed three inches of wire. His face is red with the tugging and finally he wipes it clean by stopping and cursing at the wire. I try to help him then, but we can't both get our hands on the three inches of wire, and it refuses to be pushed through from the other side.

"This isn't going to work," Mark cries out. "We've got a whole house left to do and we can't get past the first corner. That damn little book doesn't even mention anything like this."

I can tell he is exasperated. He is ready to be on his way.

I suppose I could pretend that this is a common mistake, that everyone tries to wire a three-story house with one strand of wire instead of installing the electrical outlets and switch boxes first and then cutting pieces of wire to go between them. Wiring a house by this method would be similar to writing a three-hundred-page novel with one sentence. While it might be done, the strain would finally become unbearable.

After we discover our error of technique, at lunch, I buy a sack of electrical wall and ceiling boxes and bring Heather back out to the house. I have been chagrined so many times I don't even turn crimson any longer. I was born with my brain and I have done all I can to educate it, and so I view most of my shortcomings as inevitable. The inevitable requires no apology. The end of the world can hardly be considered my fault. We walk through the house nailing up the boxes, remembering my mother's admonition, "You can never have enough outlets." At first we place the boxes along the floor underneath the windows, because we know we'll never put a piece of furniture in front of a window, blocking off the view and outlet. But after we've gone through the house doing this, Heather is unsure.

"Now you'll always see the ugly ol' outlet," she says. So Mark and I go back through the house prying up the boxes and moving them a couple of feet over, trying to find some happy distance between expanse of wall and window. We nail up the switch boxes a bit lower than usual, in consideration of Heather's height and our future children. Heather says she couldn't turn on a light till she was ten.

Once we've installed the boxes, the wiring goes much faster. All we have to do is remember which end the electricity comes from. We thread from pantry to ceiling along a joist and then up or down into walls, letting the end of the wire stick six inches out of the boxes. We use miles of wire, it seems. One breaker can carry seven or eight outlets, if you don't have a huge TV plugged into each one of them. It takes twenty feet of wire just to get up to the third floor, before we even begin to reach the outlets along the fifty-foot walls. We wire for light fixtures, ceiling fans, sconces and porch lights. We wire for heating and air conditioning, using heavy, thick black wire that's just loaded with pennies. The house is filled with tiny curled shavings from the drill and paddle bit. The bit makes a neat, round, hot hole that's nearly impossible not to put your finger in once you've drilled it.

The blond newness of the spruce that the drill reveals reminds me how long we've been working on this house. Most of the studs have turned to an ashen color with streaks of soot. It's been five months since we nailed up the greater number of them. Five months. I thought I'd be finished in a year. There's little hope of that at the present pace.

"Work faster," I yell at Mark.

When we've finished dragging wire through the house we gather two sets of pliers, screwdrivers and wire strippers and begin to install the switches and outlets. These are simple jobs usually. But occasionally, when four or five wires meet in the same small box, you finds yourself running across the room, leaping into the air and driving both feet at the receptacle to get everything crammed into the box. While electrical wire is very easy to bend when you have a three-foot section in your hands, turning a two-inch section demands inconceivable force. These efforts can tax you physically, but they require little mental work. Until you come up against your first three-way switch.

A three-way switch is necessary when you want to be able to turn off and on one or more lights from two separate places. The wiring diagram in our little book looks like the pubic hair of a gorilla. We require these switch set-ups on our stairs so we can turn the light on going down and turn it off once we're there, and vice-versa. Mark and I lay the book on the concrete floor and kneel before it in fear. We trace all the lines with one finger on the line and one finger in our mouths. We hold the various electrical parts up in the air, turning them over to see what is between their legs. Once or twice we say, "Oh, I get it," but neither one of us moves. Finally we both hit upon the idea that we don't have enough parts, that we need an extra wire. Once we obtain a spool of green wire we're able to continue, popping the switches into place and letting the light-fixture wires dangle from the ceiling until the sheetrock is installed. It takes Mark and me almost all of his remaining two weeks to finish wiring the house. My father has yet to install the breaker box in the pantry and hook to it the great gaggle of wiring hanging from the joists. After he does, I can install a meter head on the outside wall of the house and the electric co-op can bring us light.

But in the meantime, I have Mark for one last day. I have been planning this day's effort for some time but still have no plan. We meet at the lumberyard at first light.

"What?" Mark asks. "What is it?"

I bend my finger and he follows.

We walk to the far end of the long, old saw building, a building constructed of salvaged materials and used until lately to cut forming stakes. The building is open at one end. At the closed end is lodged a great green monster from some aircraft hangar or experimental wind tunnel: a steel-framed, electric-powered fan five feet in diameter. I point up to it and say, "We're going to put this fan in the third story ceiling."

Mark gasps as if I've just said we're going to kill all the puppies in the neighborhood. He gasps because he knows I'm serious, because the fan and motor must weigh three hundred pounds and because he realizes he may never see Utah.

I flip the motor on one last time before we cut the power. The fan turns slowly at first, churning, like a B-36 at the end of the runway. But then, as the motor grooves, realizes its life in the shaft and bearing, Mark and I have to lean forward into the power, like dogs leaning out the car window. Our foreheads are revealed, hair streaming back, rivers run from the corners of our eyes and our lips tremble into that famous smile of the rapid acceleration test pilot, the guy they strap to a railroad tram before, completely behind his back, they fire off a rocket. I fight my way back to the switch and cut the power.

"You'll blow all the windows out of your house," Mark says.

"No," I say. "We'll reverse the fan and pull fresh air in on the first floor and push it out of the vents in the attic. Continual circulation at one tenth the cost of air conditioning."

"But how will we get it up there?"

I summon all my bravado and assure, "One step at a time, Mark. I've got it all figured out."

"Tell me the last step now," he says.

I don't expect this, so I squint and in a Charlie Chan accent reply, "Last step always come last. But in meantime wise to watch out for ignorant toes."

Mark takes this as well as you just have. So I throw the ball joints of my shoulders up to my ears. This satisfies him. He just wants to be reassured that I'm still as dumb as he is.

We begin by underestimating this fan. When we're through drawing out four lag bolts that hold the frame to the wall, we gently tip the fan forward, spreading our feet to support the weight, gently tipping, easing her out of sawdust-filled crevices, and promptly drop the entire unit the four feet to the floor. There is a massive thump and rumble under our feet. My father comes out of the hardware store, looks at us with his hands on his hips and then turns around and walks back inside. We look down at the fan. The blades are slowly rotating. Mark checks his hands to see if all his fingers are still there.

"Maybe we should try to take it apart," I suggest.

The best we can do is remove the motor from its cradle – the fan and its heavy steel frame were joined for life at the factory – but this does enable us to lift the separate units, and we duck-walk them into the back of the antique mall van. On the way out to the house we consider the two flights of stairs and the flight without stairs. We consider the possibility of the rough-cut risers cracking under the weight of me, Mark and the fan. As for getting the fan up into the ceiling of the third story – we know we won't be able to lift it there without the assistance of some lost technology.

So we consider the stairs the easy part. I cradle the motor in my arms, testing its compact dead weight, then spinning on my heels and short-stepping to the stairs, taking them one shuddering step at a time, the greatest weight of the motor wedged in the joint of my thigh and abdomen. There's no use stopping once you've started, so I shuffle and stomp all the way to the third floor, dropping the motor on the plywood some distance from where we'll be trying to install the fan. I lie down on the floor beside the motor, working up the breath to tell Mark, "Okay, now you bring up the fan."

But it's easier just to fall back down the stairs. We have to carry the fan by the outside of the frame. If you hold the inside, the blades nick your fingers as they rotate, which is remarkably unnerving. The fan is so awkward that we can only take one step at a time. We finally make a desperate attempt at sliding the bulk up the stairs, like skipping a stone on the crests of waves. We sweat and almost lose the load several times, and can't believe that God made something as heavy as this fan. When we finally break up onto the second floor we have to stop and rest, balancing the fan between us.

“We could have put this off a couple of months if you’d stayed,” I say.

“Tomorrow I’ll be free,” he replies. “I’ll be no man’s whore. Just have to get through one more day.”

“Go to hell,” I say. “Pick this thing up.”

“One and the same,” he wheezes, and we struggle up to the third story. Once there, the fan lying on its back, Mark and I lying face down, we realize the desperateness of our situation.

“I’m weak,” I say. “I’m a weak person.”

“I’m weak and I’m stupid,” Mark says.

“Me too,” I say. “I’m a coward too, though.”

“Yeah, well,” he says, “I smell bad.”

“Yes, yes you do.” And then I put forth the only idea I’ve got, short of giving up. “The pyramids,” I say. “We can build a ramp and slide the fan up.”

“Out of dirt?” he asks.

“Out of boards,” I say.

“We don’t have any slaves.”

“There’s me and you,” I say.

And Mark says, “Oh,” in that old way, caught between hope and despair, remembering that for a slave, freedom always comes tomorrow.

Using some of the two-by-eight scaffolding material, we build a sturdy frame in the attic ceiling, directly above the stairwell opening. The longest length of two-by-eight we have is sixteen feet. I cut angles on each end of two of these and nail them as a ramp from floor to frame. The fan will ride up these two boards on its extreme edges. Our plan builds as we proceed, one step ahead of us. We lay the fan over on the ramp and have an immediate success: we slide the fan three feet up the ramp. We then have an immediate failure: the two-by-eights begin to spread, bowing outward under the weight. We let the fan slide back down to the floor.

“We’ll have to push it up a foot at a time and nail a two-by-four across the ramp to hold it together,” I suggest.

Mark, the heat beginning to take its due, nods enthusiastically, as if I’ve offered him ice cream. We slide the fan up a couple of feet, and while Mark braces himself, like a stick holding up a window, I nail a board underneath the fan across the ramp.

I smile. "Only fourteen more feet of ramp to go."

Again we slide the fan a couple of feet up the ramp. Again, while Mark locks all his joints between fan and floor, I nail in a two-by-four. When I have the final nail in, Mark, under a considerable strain for a slender person, carefully unlocks his knees and elbows and explodes like a spring across the floor. I pick him up and, seeing that he needs some encouragement, say, "You did it." Sweat streams off both of us. The windows are all open up here but there's no breeze. We're working on the breeze. Twice more we push the fan up the ramp, maintaining our progress with another rung on what's becoming a very wide ladder. The top of the fan is now at the level of the third-floor ceiling and the heavy frame we've built. We'll have to lift the fan about six or eight inches, then slide it up over the lip and drop it into its hole. I climb up into the attic. Mark spreads his feet on a lower rung. There's not much room up here. I keep impaling my head on nails sticking through the roof. I find the best footing I can, balancing on a joist and part of the framing, and heave up on the fan. As I lift with everything I have, in this sprawled out quivering crouch, the fan rises high enough to clear the lip, and I yell, "PUSH." Mark pushes. I can see all of his muscles and tendons go taut. But the fan doesn't go anywhere. All its weight is on one edge and my pulling can't help. It has taken both of us pushing up to move the fan this far. I hear Mark give it the old college try. He gasps, utters an obscenity and doubles his efforts. The fan teeters, shifts left and right then left again and begins to fall through the ramp. It will fall all the way to the first floor, through both sets of stairs. With every wiry filament in his body Mark shifts the fan back onto the ramp. I let my end drop back down. The house shudders. I can't move. Both of my calves have cramped up. Mark lies on the ramp/ladder. His face is purple.

"We can't do this," I say.

"We need more people," he says.

"It's not our fault," I say.

"It's just too heavy for two people," he says.

I climb down out of the attic, and we look up at the fan, halfway to home. Both of us have our hands perched on our ribs, elbows out, as if we're about to flap away.

"I'll do this later, some evening when I can get Phil and Bobby and Dad to help me." Mark doesn't say anything, just nods, and follows me down the stairs. We still have half a day left, so we begin to pick up scraps, sweep up all the accumulated sawdust. We do this silently, holding the dustpan for each other. It's cool downstairs. We gather all the tools, ordering them neatly under the stairway, and roll up all the extension cords. We go over a few things Mark's been thinking about. He asks how I'll go about doing this and that. Then we walk around the house together, looking up.

"I'll always be proud of this," he says. "I'll bring the boys down when they're older and show them what I helped do."

"You probably hammered more nails than I did," I say.

We stand under the oaks, looking up at the house, and maybe it's this first outdoor autumn coolness coming up between the trees. I don't know.

"Are we men?" Mark asks.

"We're men," I say.

"Let's put that fan in the attic."

"You betcha," I say, and we're there under and over the fan again, my sweat dripping on Mark's head, and finally with the strain we break past some sort of sound or friction barrier and the fan begins to come toward me. For a harrowing second it halts, wavers, undecided, threatening not only to fall but to fall on Mark as well, but I pull a gut and the fan comes off its fulcrum point, and we slide and drop it into place.

"No two people have ever done anything as great as that," I say, then think a second and take it back. "I probably wasn't doing my part the first time."

Mark tromps around the third floor, looking up at the fan as an altar, and smiles. "We did it," he says, over and over.

It seems simple, then, to carry the electric motor up the ladder and mount it to the fan. I wire power to it, and at dusk, not knowing quite what to expect, we run an extension cord to the switch and stick bare wires into the cord. We stand well out from under the fan as I do this. After one or two electric arcs, which temporarily blind us, the fan turns once, twice, and then becomes a thronging blur. It's almost frightening. It is frightening. We can't tell if it's working till we notice all the dust motes in the house rising. I am most worried about vibration, but there doesn't seem to be any. A fan this powerful might shake all the nails out of the house. Mark and I creep under the suction, trusting, meekly, our framework. The suction lifts the collars on our shirts.

"What if the house breaks loose from the foundation and starts to fly?" Mark says. He isn't looking at me. He's turning slowly under the fan, index fingers up, his arms alternately rising, doing the hokey pokey.

I join him, drying my armpits, and say, "Heather's afraid it will lift her eyelids while she sleeps."

On the way home we are very satisfied with ourselves, lolling our heads and patting our knees, and I sing an old favorite of mine, "A wimoweh-o-wimoweh," the song where the lion sleeps in the jungle, and Mark sings with me, hanging his head out the window, spectacles pushed back up against his eyeballs, a free man.

Sunday morning, there's a Volvo beep outside the back door of our garage apartment. It's Mark and Tina and the boys, all of them trussed up in their Conestoga wagon.

"We're on our way," they say.

Heather and I lean into their car through the windows and pinch the boys. They're surrounded by mounds of clothing, utensils, and vacation munchies (biscuits and jerky). They're aware that something's up.

"We're going back to Kentucky first," Mark says, "to see my parents."

"I remember Kentucky," I say.

Heather and I are leaning over Mark and Tina now, but we don't pinch them. I offer once again to let them live with us. They've been living with Tina's mother. She and Mark disagree over the authorship of the solar system, and, amazingly enough, this issue still creates tension between human beings. When Mark and Tina don't take us up on our offer, we tell them to watch out for Indians and to ford rivers only when the water is low and not to stay too long in Kentucky because the Rocky Mountain winter is almost upon us. I pat the hood of the car, as if it were an ox, as they pull away.

Bare moments afterward I find something missing. I look through this void into a sea of grass. My voice doesn't bounce back to me as it has.

"I'm going to take some time off from the house too," I tell Heather as we go back inside.

"You should," she says. "We can go to the mall. You need new shirts for winter."

How quickly I'm lost, undertow, cold currents; Davy Jones' locker is a fitting booth at J. C. Penney. You have to struggle to escape drowning. It's very easy suddenly to see myself as part of the stream, moved, but unable to move. It takes me a month to bob to the surface of a sea whose beach I've never walked. I work at our business, I read books, I fiddle with an antique car I'm restoring. I don't want to go back out to the house because I know how much work there is to be done. I suppose I am tired. That's as good an excuse as any; I've permitted people to die when they've explained that they were very tired. I lose my individuality, my train of thought. My nose collapses into my face. It's not that I miss Mark and his help to the point of distraction, but that I fear being in a big house alone. This, knowing that he'd be alone, is why Thoreau built small. There's no telling what I'll do. I'm so enamored of possibilities that I am disabled by the possibility of failure. Which, as anyone knows, is nuts. I'm afraid I might have to drive a nail twice and there won't be anyone to see it, to laugh at my already given ineptness. It's hard, when some failure is inevitable, to take it alone. I always brought dead birds home to my dad, so he might stroke their feathers sadly too.

But the fear is something I get over. The hardest part of writing a book is doing it alone. The hardest part of building a house isn't the work at hand, but the prospect of the work at hand, the mountain of unpeeled potatoes, or Chinese nails, in our mind's eye. But I don't come to this philosophy by philosophy.

Heather, sleepily, rolls to me one morning and asks, point-blank, as if she's been forced to this end, "Joe, when are you going to start working on the house again?"

And although I have other plans for the day I am so surprised she's caught me that I put on my jacket, fall into my dew-soaked truck and while on my way try to recall where I left off.

In this meantime fall has come, almost secretly, like leaves blown in under the door at night. It's the middle of November and

I've been twenty-nine for a week now. Sometimes there's a dew in the morning, but usually there's just wind, stiff and dry, stripping summer, putting a graceful arch into the tall grasses. The ends of bare branches make tight, elliptical orbits. The cowbirds have gone, but high overhead, wheeling, forming up, the return of geese. Their massed calls, resonant and somehow always unheard of, somehow unearthly, repeatedly bring me. I look up, searching the blueness, and find their haphazard flocking come into an even V and am as startled as if I'd found my name etched into a rock in a creekbed. Heather and I are quite proud that they fly over our house. Gaggle after gaggle comes over, and we hear their voices come in and go out as they turn, like the signal from a distant radio station. We listen closely, and find it hard to breathe evenly.

The house is as Mark and I left it a month ago, swept clean, the tools stacked neatly under the stairs. I almost don't recognize the place. I can't tell you, among the myriad sacks of nails lined up along one wall, which nails are in which sack. I could have done this a month ago by the wrinkles in each brown bag. There has been some activity: there are strands of spider webs strung across rooms and doorways, connecting studs, not a vast network but the spinnings of a lone traveler looking for a house site. Scattered throughout are amorphous, unsettling specks that seem to have been left by some form of insect. Perhaps they are some form of insect. In the tower of the master bedroom, lying beneath a closed window, is the brilliantly colorful, dead body of a woodpecker. He must have found his way inside somehow and perished against the pane of glass, freedom clearly in sight. I pick him up, marveling at his hue, the lolling of his head, and opening the window send him on his way. I suddenly realize I've heard my footfall. It's so quiet I can hear the spruce and glass and tin popping as they grow, expanding with the heat of the morning sun. There is so much popping and creaking that I wonder how the house holds together. I wander through the rooms, my hands in my jacket pockets, looking up at the bare joists. Once again I notice the lumber has lost its freshness of cut. It is darkened and in many places water-stained: rain before the roof was on. I hold onto studs, lean against them, touch my forehead to the cool glass of a window.

Well.

Heather wants phones all over the place. I'd much rather have a dozen mailboxes. But I comply, nailing up telephone receptacle boxes, two in the kitchen and one each in the living room, master bedroom, second floor hallway and on the third floor. The place will sound like a fire hall when the phone rings. Nearby farms will think we've been robbed. Heather believes Maine and Texas have something to say to each other. Our electric company has locally generated power: its lines and our lines run back to the power plant on Eagle Mountain Lake. But when these phones are tapped into the Contel system our house will be electrically linked with the entire world. We'll be able, assuming we pay our bills, to call Paris or Zimbabwe any time we want. The possibilities for human contact are tantalizing. We're building in the woods to keep away from their sort, the human sort, but a telephone line is a thin umbilical back to the species, akin, I suppose, to yelling at someone across a pond, words over water, we're fine.

I string the pliable, pink, four-stranded wire in series to each receptacle, leaving the bare ends poking from the open boxes. I could hook up a phone now, have Contel bring Zimbabwe to the house, but then people would just be calling me to find out if I'm working, if I'm still alive. And I'd have to count the rings on my winded run to the phone to report yes, I was and yes, I am. I trust that without the phone, Zimbabwe will assume the best.

I'm trying to think of all the things that go between the walls. Heating and air conditioning ductwork, but that will have to wait for a January option check from the *Flatland Fable* movie people. Gas lines, water lines, wiring for the butler call system, a pocket door between the living and dining rooms, and finally the insulation. All of these things, excluding the air and heat system, don't cost much, but do require some time and labor, which is just as well since I have plenty of time and little enough money. Our finances for building the house will hold out only because I'm so slow. Without our cheap garage apartment we'd be living in a tent in the master bedroom by now.

My father has promised to help with the interior plumbing, stubbing the copper pipe out of the walls. He's had experience sweating joints together in his own houses, and this is one more lost art that a father will not pass on to his son. I could learn, but expediency wins out. He already knows how, and works in a great, intimidating, debris-scattered sprawl on the floor, constantly sweeping the concrete around him with his palms looking for parts and tools. Every few minutes he has to rise, one tool or elbow short, and either run to his house or the store, or send me. In the pantry as many as six or eight pipes protrude from the floor, and neither of us can remember which are hot and which cold or even where they travel to under the concrete. We pull all the insect-repelling duct tape off their open ends. Dad walks to the far side of the concrete pad, and while I lean to the floor and blow through each pipe, he holds his palm half an inch above the other end. With deductive reasoning, a Marks-a-Lot and my exhausted lungs, we label each pipe H or C, upstairs, downstairs.

As Dad works with the pressure side, copper and solder, I work with drainage, sink and toilet, shower and tub. For this I'm using plastic pipe, PVC, which can simply be cut and glued together, not unlike an airplane model. While my father fires up his propane torch, I swab a heady glue on the end of a pipe joint and plug on an elbow. The glue dries and locks almost instantly. Unfortunately, I've stuck the elbow on facing the wrong way. It's best, Dad suggests, to put the whole system together and then go back and glue it once I've seen that it fits. If I had a baseball glove in my hand I'd fling it on the ground and stomp around. But I'm grateful to have someone else knocking around the house with me. I pause every few minutes in my work to listen to my father's working. He breathes more heavily than I do but doesn't throw things as often. He's chewing tobacco and every so often walks to the back door to spit. He calls my name, shows me his progress and says he'll be back later. I nod and say thanks, reveling in all the work he's accomplished, work I won't have to do. I am only partially ashamed that I've let my dad help me tie my shoes. He seems to get some kick out of them too.

I take my saber saw and drill to the upstairs baths, mark the positions of all the drains and cut round holes in the plywood floor. I can't resist putting my eye to the holes to look below. I stub the toilet and tub and shower drains up through these holes and the sink drains through the walls. I'm able to run most of these pipes parallel to the joists and studs, tying them off securely with metal strapping, but getting a drain into the master-bath toilet means I have to cut a four-inch notch out of four two-by-twelve floor joists. My father returns, sees this and gasps.

"Joe Alan, why'd you do that?" he asks.

"It was the only way to get the pipe there," I explain, absolutely positive that he's perceived some far simpler solution.

"Are you sure?"

"Yes," I say, exasperated.

"Well, I hope the floor holds up," he says.

"Well, me too, Dad," I say, watching him look up. Then I ask, "How else?"

And he says, "I don't see any other way you could have done it. I guess it will be all right." Then he carries his small bag of forgotten parts back to his last project.

I stand there a while longer, trying to figure out if there wasn't some other way. Then I go upstairs and jump up and down on the floor over the weakened joists. They hold. I jump up and down on them some more so Dad will be sure to hear.

My father stubs out hot and cold water to each of the bathrooms and cold to each of my hallway fire boxes, then solders on caps so we can test the system with pressure. You want to know if your pipes leak before the sheetrock is nailed on. He runs a water hose from the well to the copper, and cranking the valve open slowly we hear the sharp, severe fizz of pressure screaming into the empty pipes. We've left an outdoor spigot open so most of the air can be expelled. I run from joint to joint listening for the hissing spray of a leak and find a gusher in an upstairs bathroom. Water is spraying in a powerful fan across one whole corner of the house.

"Shut it off," I yell. "Shut it off."

"I don't know how that could be," Dad says, seemingly awestruck, but he gathers his solder and propane torch and sweats the joints together again.

Again we open the valve and again I run from joint to joint, searching for water. This time there's none to be found, only the spraying of the outside faucet. But I watch each joint for moments at a time, like a cat watching a ball on the kitchen floor, sure that it will roll in just a moment. I feel like my eyeballs are going to burst with the pressure. Finally I remember that it will be at least a month before the sheetrock goes on the walls. Any leaks should surface by then. It still seems strange to have water running through the walls of my house, skeletal though they may be. It seems like a foreign substance: this liquid. I've heard pipes shudder in old houses. Water has never been easy to control. It's all I can do to get it from my palm to my mouth. I turn my back on the pipes an hour after my father has gone home for the afternoon. In the evening at the garage apartment, I jump every time Heather opens the faucet over the kitchen sink. I'm afraid that I'll wet the bed.

Over the next week, while installing a couple of our interior architectural antiques, I'm constantly conscious of the water in the pipes surrounding me. I work in the house alone, sure that I'm being snuck up on, ready to hit my head on the ten-foot ceiling if the house gets me in the back with its ice-cold water gun.

I'm working with lumber again, fresh spruce from the lumberyard. The pine pocket door we bought at a Fort Worth estate auction for two dollars needs a pocket to slide in and out of. It's a big door, four feet by seven feet, that will go between dining room and living room. We could use a regular door here, but this one's nice because it requires no swinging room, allows a larger opening between the two spaces, and slides on this neato track and set of wheels, completely disappearing into the wall. The main supporting wall is already in place. I mount a custom-made track to the wall with heavy screws and hang the door with its original wheels. I stand back, give the door a little nudge down the track, and before I can even gasp the door is in Chattanooga, rolling off the end of the track and hanging by one set of wheels down the embankment. No need for grease here. I lift the door back onto the track and push it back down in front of the doorway, letting it hang there in the breeze like a car on a mountain lift.

All that's left to do on this project is build the false wall the door will hide behind when closed, an hour's job. I use the same ten-foot studs and plates we used to construct the original walls, building the entire wall and door opening in one unit on the living-room floor. I'll just tilt the whole thing up and alongside the old wall and pocket door. It feels good to use a hammer and nail again, the noise of my knocking reasserting my claim to this place, this idea. The sawdust is pungent even in the cool air and brilliant against the grey concrete floor. My twelve penny VC's sink into the spruce, going home with a last ping and knock of completeness. I leave the last stud out so the fit won't be tight when I stand the wall up.

The wall is sixteen feet long, ten feet high, with a four-foot doorway toward one end, and maneuvering it into place is another one of those strength tests that young people alone delight in, knowing they can probably pass. I position myself at what will be the top of the wall and lift it up three feet, letting it rest on my hip while each end alternately wags and dips. Then one quick bicep-and-snap lift to my shoulders; the wall wags even more treacherously, and my legs and feet do this impromptu dance underneath the rest of my body. At this point I start walking, shoving the bottom of the new wall toward the old one and then pushing up on the entire unit, toward the ceiling. The wall doesn't go all the way over to stand on its end. It stops a foot away from where it should, pinned between floor and joists at an angle. I can't figure it out. I stand a stepladder beside it and hit the wall with my hammer. It moves a quarter of an inch but moves no more. I can't figure it out. This wall is the exact same height as the other. I go outside and get the sledgehammer. It's very hard to use a sledge while standing on an aluminum ladder, but I'm able to pound the wall an inch closer to home by hitting the plate rail a few times. But I'm still not anywhere close. I can't figure it out. It's as if this new wall is half an inch taller than the old one. I measure my lumber, making sure I didn't get studs that were cut wrong at the mill. They're just right. I look for some hump or imperfection in the concrete floor. It seems fine. And then, with the horror of Chicken Little, I realize that my house is falling. I would scream if there were anyone to hear it. In place of screaming I stand in front of my too-tall wall and fold my arms. This is the bottom floor. The whole weight of the house is resting on my six-month-old walls. The walls have compressed under the strain, bowing the old studs perhaps a thousandth of an inch outward, making my new studs an eighth of an inch too tall. I pull down my new wall, knock off the lower plate and shave a saw blade's width off the studs. After nailing the plate back on I stand the wall back up. It takes a few blows of the hammer, but the wall slides into place, forming a pocket three inches wide for our pocket door. I walk out into the field below our house, turn, hold my thumb up before my one open eye, trying to see if the house looks shorter.

Between the walls. I'm still not in any routine. It will dawn cold or wet and I won't go out to the house for days at a time, although it's constantly on my mind. I feel guilty in the same way I would if I'd let my grandmother down. During the whole two-week period before and after Christmas I don't visit the house even once: Heather and I are busy at the antique malls, and the family wants to have all the holiday gatherings at houses that are finished. Out at my folks' house I stand in the front yard for a moment before going in. Down the hill, through the bare branches, I can see the entire roof, tower and all, glistening. It seems like the weather has been wet for months on end. Mom tells me she's seen lights down there at night but finally realized they were the reflection of the moon off the windows. She says she thought every light in the house was on, even though she knew there wasn't a light in the place.

When the New Year's rush is over I'm finally able to get back out. My father has drawn me a schematic for our butler call, another two-dollar auction acquisition. It consists of a twelve-by-eighteen-by-two-inch mahogany box with a glass front, and two electric bells. The glass front is painted black but for nine clear holes behind which dangle nine little electric flags. Over the windows, in gold leaf, are the words, "Back Door, Drawing Room, Dining Room, Living Room, Bath Room, Bed Rooms 1, 2, 3, 4." At the bottom of the glass, also in gold leaf, the legend, "Andrew Bell-Perth." A Scottish butler call then, for summoning Scottish butlers. At the push of a button in one of the aforementioned rooms, a bell rings and the appropriate flag waves behind its protective glass. The thing is an Edwardian bitch to wire up, requiring eighteen separate wires strung out to the far corners of the house. And it's a conversation piece at best, although it might be nice when one of the children is sick.

I've called an air-conditioning and heating contractor for a bid, and while I wait for him I drag what seems like the millionth mile of wire through the house. The wire is delicate, rolling off spools, and I have trouble with the tiny coils. The wire knots and kinks, and inside my winter gloves my fingers are numb. I'm warmer inside the house than out, but only because I'm out of the wind. I'm caught up in my cold world, snorting ceaselessly my own snot, unknotting tangles, threading the same studs with a finer strand, when the back door opens and closes. I step downstairs to a view of his nostrils. He's walking from room to room with his face pointed up.

"Great house," he says, in a, somehow, sixties accent. Then he lowers his face at me and smiles. He's missing a front tooth (why is there such an archeological bent to my thinking?).

"It will be someday," I say.

"No, it's great now," he repeats.

I look around again. I guess I get caught up in bothersome details, like walls. I realize that this is the stage he always sees houses in, all the guts hanging out. I show him where I want the downstairs unit, and together we walk the cold concrete, tracing the path of the ductwork through the joists to each room. Heather and I have decided on two units, one downstairs and one up. He thinks this is best too. We'll be able to shut the downstairs unit off at night and the upstairs off during the day. And if either one breaks down, there will still be a warm or cool spot somewhere in the house to huddle in. The upstairs unit will be more difficult to place. It has to go in a low wing in the attic, and the ductwork must travel through the smaller two-by-eight second-floor joists. The third floor will have to go without for now. There's room for another big unit in the other wing of the attic but there's no room in our wallet.

The contractor tells me what kind of platform to construct for each unit, what kind of electrical power he'll need and where to place the propane line. Then, sitting in the cab of his pickup, the motor running, he writes me an estimate. I wait for him on the back porch, the wind running water out of my eyes. I clap my bemitted hands together and stomp the concrete while he sits in his warm truck. I expect the worst because the worst has become common as far as bids go. I stand there so long waiting for him that I begin to have bad thoughts about him, adding dollars to my bid. But when he steps out, calling me by my first name for the fiftieth time (I can't tell if that's just his way or if somebody's told him this is a good technique of salesmanship), his bid is two hundred dollars less than my father estimated the job would cost. My father is notorious for underestimating the cost of things. So I tell him okay, let's get started, and I give him a check for half. He says he'll get everything ordered and even lends me his pipe-threading kit after I tell him I'm going to run the gas lines.

When he's left I step back inside, blow on my hands, remember where I was with the butler-call wiring and decide it's too cold to work, that my quality of life would be far better at home with Heather, where the heaters already work and where she is baking bread.

Through the next week I finish the butler-call wiring, mount junction boxes for the air-conditioning units and strap a meter base to the outside of the house so the electric company can hook us up to power. I build the platforms and the tiny room for the downstairs heating unit. I buy my first sheet of wallboard, sheetrock, and line the inside of the heater closet because I won't be able to do this after the heater is in. This one sheet, with others in the past, convinces me the sheetrocking is something I want to hire out. Apart from the general distastefulness of the task, doing the entire house by myself would take me months. Even on this single piece I have to cut a six-inch-by-six-inch hole to get the electrical box in sight. Dad comes down again and wires the breaker box, promptly stripping all my identification tags off every dangling wire. He shows me what things in the box never to touch and how to replace breakers when one goes bad. I ask him which breakers go to which outlets and lights. The air-conditioning units are obvious, they're the big breakers, but everything else is a blank.

"Well," he says, "that's no problem. When the power comes you can just turn all the lights on and turn off the breakers one by one and label them."

"But I had them all tagged already," I whimper.

"I had to cut those ends off. They were too long."

"But, Dad."

"Well, what else can I do?" he asks, stepping through the house.

"Nothing. Everything's done. I just have to run the gas lines, then the contractor can put in his ducting and I can call in some sheetrockers. Oh, yeah, the insulation."

"Boy, that will be a big day. Everything will look completely different. There'll be rooms."

We nod and beam together.

"You'll be living out here before you know it," he says. "We'll build us a big metal shop between the two houses and have a wood shop and a car lift and an engine hoist and a wall of parts bins for all our nuts and bolts, everything where you can find it. All our tools on the walls."

This shop, common dream of all mankind, has been built many times in my and my father's minds. I have to wobble my head in a cartoon-like way to bring myself back to my as-yet unfinished house.

"You have to be careful with those gas lines, son. One leak and you and Heather and the house all go up together. It's not like a water leak or a breaker popping."

"I know," I say. "I know. I'll be careful."

"You'll have to test the lines under pressure before the sheetrockers come in."

"I will," I say.

So it begins. The winter is at my side. All the leaves have made their yearly migration from bough to earth, uncanny instinct. Myself, I have to reason it out. I hold a piece of pipe in my hand as if it were a tree branch. I'm surprised to find that there's a hole all the way through it. I look up in the air and think, "What could I use this for?" My first thought, of course, is that it has a good metal heft and that I could reach out and really bonk somebody on the head with it.

The winter is at my side. It's so cold I have to work quickly or not at all. The gas pipe is black, and coated with a protective oil. I have to put more oil on to cut threads in it. Strangely enough, when my hands are coated and black with the oil they're also warmer. The pipe itself is very cold, like a Coke can out of the refrigerator. I start in the laundry room, where we'll have a gas dryer. I cut each joint on the concrete floor, holding the pipe with a big pipe wrench, holding the pipe wrench with my knee. The die for threading the pipe fits into a big wrench with a ratchet. This gives me some leverage for cutting the threads but it's still difficult. I pour oil on the pipe and die, trying to make it easier on myself.

The metal shards come off the black pipe in razor-sharp curls, gleaming in the oil. Then I cut another piece of pipe, thread both ends, screw it to the last fitting, elbow or tee. I work from the dryer to the heater to the water heater to the kitchen stove, then upstairs to each bathroom and finally to the attic and the other furnace. I rub the threads of the pipes with a putty stick before joining them, to help stop any leaks. The system takes three entire days to complete. At the end of each day I'm black from head to toe with the oil, and my threading arm hangs limply to my ankle. I drive home literally with one arm.

On the evening of the third day, after capping off all openings, I screw a pressure gauge onto the end of the last pipe and pump thirty-five pounds of air pressure into the system. Propane is usually at only five or six pounds of pressure, but I want to test the system well. If the system still has thirty-five pounds of air in the morning, I can be fairly sure it's safe for gas.

In the morning the needle on the pressure gauge has dropped from thirty-five to zero. This is not good. A comparable leak sank the Titanic. I fill the system with air again and, using a thin soapy solution in a squeeze bottle, check for leaks at each joint. I find several. A leak blows bubbles. I tighten the joint with a pipe wrench till the leak doesn't show and then check the other end of the pipe. The problem here, since the threads on both ends of a pipe are turned the same direction, is that tightening one end loosens the other. I tighten one end, get down from my ladder and move it to the other end, squeeze soapy water over the joint and let it run down my arm inside my coat to my elbow, and then tighten this joint. I move my ladder back and forth a dozen times, trying to plug leaks at both ends. Finally, after hours of this, all the leaks seem to be checked. I fill the system to thirty-five pounds again and go home.

On the morning of the fifth day the needle is again at zero. I am at a loss. I stomp around the house on the cold concrete and utter niceties. I fill the hated soapy water bottle again, and again check for leaks. They're everywhere. All the leaks I worked on yesterday have returned. I stomp around the house some more, the elbows of my coat cold and wet, trying to deny the conclusion I've already reached. I'll have to take the entire system apart and retighten the joints one by one. This takes all day. In the the late evening I fill the pipes with thirty-five pounds of air again, curse the needle and drive home.

On the morning of the sixth day the pressure gauge stands solidly at zero. All the wind has left my pipes. I stand in front of the gauge, my hands in my coat pockets, dumbfounded. I think, this is as far as I can go. The house has beaten me. My throat tightens and I want to cry. Instead I pick up the smaller of my two pipe wrenches and heave it, ricocheting off studs, through the house. My mind is numb with self-pity and anger.

On the bleachers outside, James T. Feathers and his wife have taken a top-row seat. He looks at his hands, shaking his head slowly. He knows I've lost my patience.

At the hardware store everyone shakes their heads. Dad suggests the die I've used to cut the threads is no good, that it's leaving a jagged cut that the air escapes through.

"Did you tighten everything?"

"Of course," I say.

"You didn't over-tighten anything, did you? You can do that too, split the pipe."

"I don't think so," I say, more slowly.

"Take one section of pipe, thread it, cap one end and then put your gauge on the other end and see if it holds pressure. You might have defective pipe."

This seems like a brilliant possibility to me, a loophole for pride. I race back out to the house to give it a chance, threading a four-foot section of pipe, capping it off and filling it with air. I take the section home to my desk so I can watch it more closely. Oddly, I'm hoping the pipe fails the test so that I'll be redeemed. I've become as thin as the air in the pipe, under as much pressure to escape. I watch the thin black needle of the air gauge with my lips tightly pursed. I check the pipe at hourly intervals throughout the day. It doesn't move. The last thing I do before bed is check the gauge. It hasn't moved. On the morning of the seventh day, even before I pee, I check the gauge. Thirty-five pounds. The only method I know to relieve the pressure is to make the problem someone else's. I call a plumber licensed to install gas lines. He meets me at the house in the afternoon. The pressure is gone now. I feel like a collapsed balloon.

I'm standing in my own house with the plumber and the pipes, this roadblock to progress, hoping he won't be able to figure out the problem either. I am so shallow you could step in me and not splatter the fellow next to you. I stand here telling him all I've done, yanking my tail up between my legs, splitting my nuts, yanking so hard my feet leave the ground in little hops. I've asked him for an estimate to repair my mess. And I don't know if it's out of genuine goodness of heart or revulsion at dealing with an amateur's work, but he tells me I don't need him. The only problems he can see are that I didn't thread the ends of each pipe quite far enough and that I used putty rather than Teflon tape to seal the joints.

"But," I say, "here's the putty stick I used. It says right on it 'For Gas.' "

He doesn't even take the putty that I'm holding out to him, but says, "Eggs aren't good for you either but your grandma told you they were. Take the whole system apart, tap another half inch of threads on the end of each pipe and use the Teflon tape, about four wraps to each set of threads."

I offer to pay him for the call, but he won't accept anything, wishing me good luck. When he's gone I drop the putty stick and its great big lie on the concrete floor in the middle of a big room and jump up and down on it till it's saucer-sized, thin as gas.

Through the eighth and ninth days I rebuild the system again, threading the ends of each section of pipe a bit further and then wrapping the threads with white tape before screwing them into the joint. I work with an uncommon deliberateness, as if I'm stalking a rhinoceros. When, on the tenth morning, I find that the gauge is still at thirty-five pounds, that the system is tight and completed, all I can do is stare in pity at the needle, having vanquished it. The most I can muster is a patently false smirk and snort, man triumphing over nature, as if it had a chance. Almost immediately I forget the arduous self-doubt and pain and grease of the past nine days. The human ego is as massive as the future.

The heat and air boys can come in now. I have a 250-gallon propane tank set outside the library, and the power company finally runs a line from their pole to the house. Electricity is now as close as the many outlets. While the heating units are being installed I work at the other last step of what goes between the walls: insulation. While this requires the least technical know-how, it extracts the greatest misery. Rolled fiberglass insulation, as no other building material, becomes part of one's being. Even though I wear a long-sleeved shirt, gloves and dust mask, my skin fairly glows with the delicate introduction of glass fibers. Even though I try to avoid direct contact, the ten-foot-long strips inevitably lounge across my head, draping me to my shoes. The gloves prove to be such a bother and failure that I eventually discard them. The mask is effective in keeping out the insulation fibers, but it makes my nose hot and fogs my glasses. I get rid of that too. Heather comes out in the late morning to watch the air conditioners being set in, and I enlist her. She unrolls the insulation and cuts it to length while I staple it between the studs. The inside of the house comes to look like the inside of a paper bag. With each wall we finish, we stand back and try to ascertain the change in temperature. Then we curse the fibers again, rubbing them deeper into our skin.

The heating units and all the ductwork are in place within a week. I go over everything with the contractor and I'm glad I didn't try to do it myself. It would have taken me years. Perhaps my experience with the gas lines has chastened me more than I originally thought. He tells me how to keep the units up to par, how to light the pilot lights, to change the filters often. The only thing I don't like is the big ugly vent pipes sticking out through the side of the house. They're necessary to vent the gas heaters, and I didn't want them going through the tin roof, but man, they're ugly.

It takes Heather and me another week to finish the insulation. The first and second floors need insulation in the walls only, but the third has a great vaulted ceiling. I climb up the ladder with a long length of fiberglass draped over my head and fight it into place—something like trying to force a live snake into a bottle. Heather shouts encouragement from below. She wants to be cool in the summer, warm in the winter. We finish up in the exposed third story of the tower, cutting the insulation into thin strips and jamming it between the studs. Then, with half a roll left over, we pull the pink fiberglass off the paper backing and walk through the house with it, stuffing the cotton candy into cracks and holes, anyplace where cold may try to shiver in.

In the evening I call sheetrockers, since we're now ready for them. Between calls I try to remember anything I might have forgotten. I've forgotten the dryer vent. I'll have to put that in. And in between these forgettings and rememberings, while I'm sitting in front of the phone, it rings. A phone is very loud when you're sitting right over it, thinking about picking it up. I pick it up and someone says, "Hello, Joe," and I'm at a loss for a moment, but then I know it's Tina, and I say hello back. We haven't heard from her or Mark since they left. She wants to talk to Heather, so I hand her the receiver and stand there while they talk.

"Wyoming? No, it's not finished yet. We're ready to put the sheetrock in. Wyoming? Well, Joe didn't work for a while, and then the gas lines took a long time. He's working in a hardware store? Well, why did you stop there?"

Heather hangs up the phone after a while, and I wait for her to get her thoughts organized, but all she can do is look up at me quizzically and say, "You're not going to believe it."

I suppose it's hard for any of us to believe that a whole world exists beyond our personal experience and vision. A truly devout person is the most amazing thing I've ever come up against. I'm surprised by every highway that doesn't end over the next hill. I was quite sure that Mark and Tina died when they pulled out of our yard. But they're alive, in Wyoming, vowing never to leave it, never to leave heaven, for mother, Jesus or even the U.S. government. They pulled into a little town they'd never seen before and decided to live the rest of their lives in it. Mark is working in a hardware store. Tina's working in a Shakey's Pizza. Adventure is the day as you live it. I walk over to my collection of pocket compasses and make sure they're all still pointing north.

27

The explicit authenticity of my past. I believe it is real. Many stutterers don't stutter when they talk to dogs. I speak to this paper as if it were a dog, head tilted, ears perked, compassionate, unable to comprehend. I jump when it barks.

I lived for a year in a small concrete-block room in Holmes Hall at the University of Kentucky and spent another year in a slightly larger concrete-block room at Farmhouse Fraternity. Concrete block is the precise building-material match to college undergraduates. Combine block with concrete floor, plaster ceiling and steel-clad windows and you have a virtually indestructible dwelling. This type of building may be sandblasted, washed down with a hose and painted anew, removing all traces of former occupants. I, for one, felt I was probably the first human to live there.

I read *Walden* that first fall of college, and began keeping a horrid philosophical journal, and was electrified in some way by all the possibilities of knowledge and solitude. Wasn't I lonely already? Wasn't I living in a room the size of Thoreau's hut? I copied some of Thoreau's lines into my journal: "I have a great deal of company in my house; especially in the morning when nobody calls . . . . to stand on the meeting of two eternities, the past and future, which is precisely the present moment . . . A man thinking or working is always alone, let him be where he will . . . . It is a faint intimation, yet so are the first streaks of morning." This last was my favorite because I thought it spoke not to me but of me. I was eighteen, only an intimation of what I might become, with the possibility of attaining a whole day itself, something unimaginable compared to my present, confused self. I joined the Thoreau Society. I'd finally found an acceptable religion, based more on my imagination than on transcendentalism, more on my perceptions than my knowledge. It was a sort of selfish religion, focused on self-confidence and self-esteem, but it made me study hard.

In the meantime I joined a fraternity—an act diametrically opposed to my individualistic theology, but my friends were joining, and there were girls around the chapter house all the time.

In the spring of my freshman year I took my first college English class. Our class met in the ROTC building, which also contained a room with a basketball hoop for the cadets. I think their games helped us vary our sentence structures and lengths, quick dribbles, the rhythm of a fast break, the airy pause of the ball on the way to the hoop. By waiting for the second semester I had joined a classroom of kids who'd failed the class the first semester, or who, hating English, had put it off. We wrote essays for Razak Dahmane, a teaching assistant trying to finish his Dickens thesis. He met our challenge with vigor and humor and with mesmerizing concern. He laughed with great abandon. But we were demoralizing. How he woke up to us, an eight A.M. class of failures and fearfuls who were themselves barely dressed and awake, still confounds me. I felt sorry for him because he had to deal with us, and although I wouldn't help him any in class by speaking up, I wrote and wrote and wrote for him, thrashing my mind for something original. I wanted his queer laughter directed my way. I wanted his rumpled corduroy jacket with the leather patches on the elbows. I wrote my classification essay on the subject of toothpicks, an idea I'd borrowed and elaborated upon. He read it to the class, laughing so much as he read that I was embarrassed in front of my struggling peers. Embarrassed, but I'd have it every five minutes if I could get it. I'd be the lab rat who neglected food and even sex to overdose on writing.

During sophomore year most of my classes were business oriented: accounting, statistics. Worst of all was Business English, a course where we were taught language as pickpocket. We wrote letters emphasizing the importance of "You," the customer, in order that we, "your servant," might lower your estate and increase ours. For the first time in my life I hated school. I slogged to class and sat in back rows, lids half closed, creating great notebooks of doodles. At the end of my sophomore year I did two things I'd never done before and have yet to do since. I ripped up a report card and burned a book. In the fireplace of the chapter house, Mark Laughlin and I both burned our copies of *Business Account-*

ing. It made good business sense: the bookstore wouldn't buy them back, and the burning of them created heat.

Perhaps I wouldn't have been so antagonistic towards accounting if I hadn't taken Ed McClanahan's creative writing class the same spring. I'd walk across campus from accounting to McClanahan and undergo a complete chemical change. My brain, which had been floating in syrup, began to boil in a solution similar to battery acid. My jeans crawled right up the crack of my ass. I couldn't keep my fingers out of my face. Great chunks of matter broke loose from the corners of my eyes. I held the few chaste, white, typewritten pages in my hands as some sort of offering. I felt as ecstatic as the soon to be saved, the soon to be rich, the soon to be free, the soon to be screwed. I arrived in class, sat down alone, ten minutes early, quivering in fear.

During my freshman year I thought of ways to spend the weekends at the University of Kentucky; during my sophomore year I found myself packed and on the road home every Friday afternoon. I wasn't ever sure about what I'd do once I got home but I wanted to be there. My father, after seventeen years with IBM, had decided to quit the endless round of meetings and open his own business in Texas. Sally and my mother left for Texas in January while Phil finished high school, Dad closed out with IBM, and I wound up my sophomore classes. Our last duty was to clean up the farm and auction it and the house and equipment off. I think we all sensed, my father foremost, that this was a true passing, a marker in our family's life. The farm was the first "place" my father had ever owned. It had an identity, a look, to which we'd all contributed. We felt for the first time rooted to a piece of ground that grew grass and to all the past of that place, all the people who'd lived there before us. Strangely, we were unhappy with the idea of other people living there after us. Every family who came to look the place over with thoughts of buying was outlandishly unsatisfactory. We made fun of their looks after they'd gone, chastised their apparent stupidity, their obvious inabilities. Zeke brazenly sniffed at their crotches, backing up awkwardly afterwards and snorting. I think that perhaps only the television Waltons would have been good enough for us.

I said my goodbyes at Farmhouse at the end of the semester, but they weren't nearly as long and broken as the goodbye to the farm. I knew the friendships would last over time and distance, but I thought the farm was gone for good. Zeke and Phil and my father and I walked over it together the last time, down the board fence to the creek and along the creek to the corner fence and then back up over the three hills to the house. Zeke looked for rabbits and we looked for things we might forget. We looked for things so familiar to us that we immediately sensed their loss. We were afraid that if we slapped our pockets they'd not only be empty but we'd discover our hands too were gone. My father said there were things he was going to do had we stayed longer. So there was not only a past lost in our leaving, but a future too.

28

The big cover-up begins. I've ordered five hundred sheets of wallboard, the heaviest substance known to man, and it arrives on a semi. Two guys my height but thick as oil drums unload it two sheets at a time, making stacks in the center of each room. They even carry it up to the second floor but by contract go no further. Phil and I carry seventy-five sheets to the third floor and become thoroughly ashamed of our physiques. Any doubts I'd had about hiring a sheetrock crew were choked to death by these seventy-five sheets. I realize that the two men unloading the truck do this kind of work five days a week. The only job that could demand more pure physical strain would be nailing the stuff to ceilings. I await the crew, expecting seven or eight semipro football players. The old man who'd come out and bid the job was anxious for it, told me so up front, and said he could have it done quickly, would even sweep out the place when his crew finished. He wore overalls and his beard was white, and he walked with the same kind of limp my grandfather had. His bid was ridiculously low, and so I told him to send his crew on. Building is slow in the winter, and slow in

Texas now anyway. I had my pick of crews at my price. I finally felt myself a true contractor, supervising a first-rate job.

The old man arrives first, in a twenty-year-old pickup. A young man rides with him, perhaps a year out of high school. I show them the boxes of nails and where I want the special water-resistant sheetrock to be placed: behind the sinks, tubs and shower stalls. I have to leave soon so I ask the old man to make sure the rest of the crew knows where the special sheets go; they're expensive and I don't want them winding up in the living room. He looks at me as if he hasn't heard me and so I say it again, a bit louder.

"I heard ya," he says. "The crew's all here," he says. "Mike here was right behind ya when you told me."

I look at Mike. I look back at the old man.

He says, "We'll get your sheetrock up for you, mister."

Anything I could say would invariably insult them. I had been thinking that this job might take a couple of days, but these guys are going to have to move in. The most insightful thing I can come up with to say is, "Oh." I leave them setting up ladders, filling their aprons with nails. Another interminable step on the road to completion. I always root for the underdogs, but I never really think they've got a chance.

When I walk out, even though my back is to the bleachers, I raise my arms and drop them slackly to my side in exasperation. Then I think I notice out of the corner of my eye something huge and black in the underbrush. But there's nothing moving now, perhaps it's just me. I feel stupid for a moment, having raised my hopes vainly that way; still, the warmth and power of the moment go with me.

In the evening Heather and I return to check the crew's progress. We walk in the back door, stepping over oddly shaped scraps, and find our kitchen huge. It's completely sheetrocked, and no longer can we see through its walls to the rest of the house. We were sure that each room would seem smaller. Instead we find that closing off the view makes each room become a whole world of its own. The crew has left for the day, and so I don't have to check their work in a surreptitious, glancing manner. Immediately the quality of the work is evident: tight joints, ceilings done first, electrical outlets cut out cleanly. There is no more or less done than I thought there'd be. The crew is doing the job by the piece, on their time, so I'm happy with the workmanship.

Over the next two weeks I get a chance to watch the old man and Mike work. It would be impolite just to stand there and stare at them, so I busy myself by backing the little truck up to the front porch and filling the rear of it with sheetrock scraps. Mike is the old man's nephew. They work most of the day with no word between them but numbers, measurements and occasionally a patient "Wait while I get this nail in." They make the ceiling work look almost easy, climbing up on the sawhorses, positioning a sheet and then holding it in place with their heads while they anchor it with a few nails. Mike, who wears a baseball cap, tells me he has to be careful of the button on top. If it comes between the sheetrock and his skull the pain is intense and his knees buckle.

The house is soon deep in a fine white dust. It fills every pore in the floors and even hangs thickly on the window panes. The bare studs are quickly disappearing, and that old fear of having forgotten something overcomes me. I wonder when I will outgrow the sensation of having gotten all the way to school with only my underwear on. The sheetrock has turned our barn into a house, stalls into rooms. If there was ever a time to put a skeleton in a closet, this is it. These beams won't see the light of day till the house's doomsday. Heather and I can't think of any personal skeletons, but we do throw a few current magazines between the walls so that when the house is someday repaired or torn down the wrecking crew can pause and have a good read on us. Heather wants to throw in my old *Playboy* but I ask her, "What kind of people do you want them to think we are? Besides, I haven't finished reading it yet."

"It's four years old," she says. And plunk, before I can make a stabbing dive, she drops it eight feet down into the wall.

I feel the loss already, envying the lucky bastard who tears down my house. But then I realize that might not come to pass for another hundred years, and a hundred-year-old girly magazine is no prize in any age. I try to imagine what a girly magazine of the future will be like, but my imagination falls short.

It takes my sheetrocking crew a full two weeks to reach the ceiling of the tower room, and if the quality of their workmanship falls off toward the end I cannot blame them. At least they are talking more to each other by the end of the job. The old man confesses that it's been a while since he's "hung any rock" and that he almost died one day the week before, could feel his heart breaking out of his ribs. I tell him I would have been mad as hell if he'd died in my house first, having reserved that right to myself. After I've uttered this I realize I've made some sort of confession. Oddly, my body has the same physical reaction it has when I tell somebody I love them for the first time. I'm aquiver, can't stop the chilly knocking of my knees.

The old man says, "This is a fine house, young man."

And I nod, my hands in my pockets, almost blushing. I write Mike and the old man a check for a hundred dollars more than they bid, and thank them for the good work. I tell the old man, if it's all right, my wife and I will sweep up, and he says, sure, he's not the best of sweepers anyway.

Heather and I are. We've swept this house so many times it's like scratching the back of our dogs. Both of us benefit. We throw the bigger scraps into the truck and fill an eroded ditch with them, then back up through the field to the house again, where we sweep the chalky dust till we become members of the seventeenth-century French aristocracy. Our whole bodies are heavily powdered and we move through the hallways like ghosts.

I'm sweeping out the dining room, chalk motes so thick I can barely make out the end of my broom, when I look out the window and see quite clearly a big black dog, that ebony sniffer, sitting on the bleachers. I smile. It almost brings tears to my eyes, and I want to call Heather but she never knew Zeke and might not see him. He pirouettes on the top bleacher and lies down in a tight curl, his big, greying snout resting on the pine. I've watched him do this a thousand times, and know he'll stay there till I'm finished.

The following week we hire another crew. This one is only half as large as the sheetrocking company. Reggie is a tape and bedding man, a painter. He's working for a big construction firm on a housing development but says if we're not in a hurry he can do our house evenings and weekends. We want him because his price is reasonable and he's supposed to be very good. Our house will be difficult to tape and bed, because we aren't going to texture or spackle the walls and ceilings. We want smooth walls that we can paper later. Reggie will try to make the sheetrock joints as flush as possible. It will require an artist's hand.

Over the next month, into mid-March, Heather and I have little to do at the house. We come out every other day to see if Reggie's done anything more, walking through the rooms slowly, looking for wet mud. It seems to be an everlasting process. Coat after coat of mud goes on. Each layer must be allowed to dry, then it's sanded, and Reggie applies another layer, "floating" the tape line out into the wall. The nailheads must be filled several times also. The mud shrinks as it dries, and although it's flush and perfectly smooth when wet, when it's dry you can see the slight indentation across the house. There are problems with the weather for a while: the mud can't be worked when it's freezing. But our greatest concern is that "not in a hurry" means something different to Reggie than it does to us. He works diligently for a week and then we don't see him for days at a time. He comes out and works half of a Saturday and is gone for the rest of the weekend. It's my fault: I should have hired a crew at three times the money and had the whole house done in a week. As it is, only the downstairs is finished after three weeks. At this rate the house will take another month and a half to finish taping and bedding and then who knows how long to paint. So we stop Reggie on the second floor and turn him into a painter, have him rent a spraying rig and paint the entire first floor so Heather and I can begin laying flooring when he moves on to finish the upper stories. With the spraying rig and an inexhaustible supply of five-gallon drums of white paint, he paints the downstairs in a couple of evenings.

Heather and I pull the covering paper off all the windows and electrical outlets on a Saturday morning in late March. The light streams in, some of the first warmth of the season, and we're almost blinded by the interior of our house. The light is reflected from every ten-foot-high white wall and ceiling. The floors are white with overspray. We walk from room to room, light following us, whiteness, whiteness, brilliant with possibilities, all the blank freshness of an inevitable future suddenly before us.

29

Most people wait a lifetime to go back home; I did it in ten years, back to the streets of Saginaw, Texas. All of our old forts were gone though, the empty fields full of new subdivisions, Greene's Grocery a vacant concrete block shell. My grandparents had moved from the lake to one of the subdivisions. Mom and Sally had been living with them for six months, and Phil and Dad and I moved in too while we looked for a house. Texas welcomed us back with open arms, some forty straight days of heat of a hundred degrees and above. My father was anxious to find a new house, but my mother and the rest of us were as at home as we'd ever be, lodged in front of my grandmother's air-conditioning vents. My grandparents had given us refuge countless times, when my father was on foreign duty with the Air Force, whenever he'd have to go to an IBM conference or another overnight meeting. We knew the glory of my grandmother's food, my grandfather's ready laugh. My mother was as happy as I'd ever seen her, all of us back together, in the daily life of her parents and brothers and sisters. Mom is a cold-blooded creature, and the warmth off the pavement suits her. She'd almost frozen to death in New Jersey and Kentucky.

Mom and Dad closed on a house soon, a four-bedroom, blond-brick ranch in Lake Country Estates on Eagle Mountain Lake, only a few miles from Grandma's house. They bought the house at some $30,000 under the market price, thanks to the previous owner's taste in wallpaper and carpet. When I first toured the house, Mom warned me not to look at the walls or floors, leaving only the ceilings open to my view. I couldn't help myself. I've never had the opportunity to see a murder victim, but I don't think I'd go far out of my way to miss stepping over the body. I looked at the carpet, the walls, only a glance, and turned into a pillar of salt. The entire house was done in bright, almost fluorescent, blues and greens, brought more clearly into focus by walls of mirrors and foil-backed wallpapers. The carpet, dog-back shag, intertwined ropes of army green and sky blue. The interior woodwork, fine oak, had been stained blond to match the exterior brick. The saving grace of these features was that they were all easily covered or removed. This we did promptly, repapering, replacing carpet, painting the walls. To my great dismay, my parents did not remove the floor-to-ceiling wall of mirrors in their bedroom. For almost four years I walked into that bedroom and shied from my double on the opposite wall, the one who understood my parents better than I did.

My bedroom, originally a bright yellow, I painted a dull beige to suffer my scholarly attitude. We'd sold a great deal of our furniture at the farm auction and so my mother bought me an English oak bureau/desk for $225. My father, trying to start his own business and flabbergasted at the closing costs on the new house, was appalled at the price. I felt a bit guilty at the time, knowing full well that I lusted after the piece of furniture for the desk half alone, the worn oak ink-blotched drop front, the pigeonholes behind; I knew how improbable it was that writing would ever pay for it. My mother argued it was really two pieces of furniture for the price of one: I could do homework on the desk and keep my clothes below in the bureau. I work at this desk now, the drawers below crammed with manuscripts, reviews and clippings, but I've yet to pay my mother back in any monetary way.

My decision to change majors from Business to English hardly surfaced. My parents shrugged and agreed, you had to do what made you happy. I'd arrived too late in the year to enroll at Southern Methodist University, my school of choice. SMU was the only university in the area offering a major in creative writing. So during the fall semester I took two night classes at the University of Texas at Arlington: astronomy and creative writing. I worked at Pinetree (my father's business) during the day, soldering connections and checking circuit boards.

I sat up till all hours in the new house, at times talking over the latest of my brother and sister with my mother, or watching some long-dead movie with my Dad. We'd eat cereal or popcorn. But usually I sat up by myself, at my desk or at the dining room table (closer to the refrigerator), and worked on stories so dense I can't now follow them. It took me almost a year to realize I was more unsatisfied with these stories than my characters were with their lives. And this is a Statement, because these people were killing themselves, killing each other, going insane, apologizing, on and on. The devil visited their houses.

I think, perhaps, he visited mine, our, house, too. It's childish really, blaming a house for a time of life, coincidences. Good things happened there too. But I hold little fondness for those rooms. We all had to compromise too much there. It was the first and only house I've ever lived in that never came to be a home. Old Zeke died there, dog of my childhood, great black immortal hound, unbearable sweetness; my palm yet cups his cool, velvet ear.

30

Thoreau never says what he walked on, what kind of floor kept him out of his cellar. The shanty he bought for materials had a dirt floor, accommodating chickens as well as humans. I think this kind of floor would have gone well with *Walden*'s theme, a man's feet planted on the earth, but practicality suggests that moles might have gotten into the manuscript had there not been a pine floor for them to knock their heads against. A roof may be a necessity but a floor is pure luxury.

No decision has tested our marriage more than this: what kind of flooring we'll walk on from bedroom to bath to kitchen. I never knew our feet were so sensitive: carpet, tile, linoleum, wood, the variety of style, composition and color, every point a possible divorce. I suggest leaving the concrete slab and three-quarter-inch plywood, using them for flooring. They're durable and can be painted any color we desire. Heather's eyes narrow to razor blades and she flings them at me, shaving off all the hair around my mouth.

"Okay, we'll do it your way," I say.

She chooses, for the kitchen, a vinyl tile with white background and silver/grey geometric designs. We'll be able to spot broken eggs very easily. The directions on the box of tile suggest repeatedly that the surface be free of all dust. I look at our floor: a layer of sheetrock dust glued to the concrete with a coat of white paint, frequent splotches of taping mud and many tracks of the canine. I wouldn't be surprised to discover spoor of rhinoceros in the corner. We find that sweeping is useless. What dust does come loose settles quickly back to the floor. Heather brings in a water hose and I beat a hoe into a scraper. The paint is water-based, and after two hours of scraping and brushing we've collected a bucket of grey goop and spread the paint more evenly across the kitchen floor. When the water dries we begin laying the tile by popping an extremely dusty chalk line along one wall. The box advises popping this line out in the middle of the room but I don't see any logic in this. I'd probably have to cut more tile and these things cost a fortune. (We've had to borrow $20,000, a debt of $452 a month for five years, to lay this flooring and the rest, to make the house livable by summer. We're at a point where we can spend money as fast as we could wallpaper with it. It costs more to floor this house than to insulate and heat it.) The glue for the vinyl tile comes in a great beast of a bucket, a five-gallon tub so heavy I have to carry it with both hands suspended like a pendulum between my bowed legs. After I've cut the bucket lid off with a hatchet I test the thick, mustard-colored paste with my finger. It's nasty all right. My steel trowel has four notched sides and my tub-of-glue's directions say to use the eighth-inch size. I've never laid any tile before, but I can't hesitate because Heather's watching me with her hands spread on her thighs. She's looking for the least excuse to jump in and take over. We're both first-born children and the competition can be fierce. We're both experts at snatching something from somebody's hand and showing them just how to do it. I dip my trowel into the goop and with a great display of dexterity and professionalism submerge the instrument in the glue. This would be a minor slip except for the fact that my hand is still attached to the trowel. All Heather can say is, "Eewwww."

"Here," I say, holding the great yellow irradiated glop of my trowel and hand at her, "you take over."

I keep waiting for the wisdom of experience to kick in, but it doesn't come. I have to learn how to breathe over and over again.

I scrape as much glue back into the bucket as I can and use a quart of thinner to clean my hand.

Taking more care, I spread the glue in eighth-inch-high ridges, enough to lay down half a dozen tile. The first tile is critical. Its lines radiate across the planet. By the time the first row of tile reaches the far wall it could be in the middle of the room or even outside. I press the tile down, adjust it a fraction, clean up a glue smear, and Heather hands me another tile. We lay five more tile and see quite readily that we're off. Each tile has to be moved a bit. The glue is still wet, so we slide each tile back a sixteenth of an inch. Then more glue down, more tile. We count the remainder of our lives in rows instead of years. It becomes knee-bruising work. The concrete doesn't give; our blood vessels do, but what's worse than the work is the standing up at the end of each row. Please, let me stay on my knees. We try kneepads, but the straps cut into the backs of our legs. We assuage our pain with the shining row. We realize that this floor is the first really finished part of our house. It gleams in all its no-wax glory, line upon line, never more than a mop of care. The bucket of glue lightens as we tile on, the geometrics allow fewer mistakes. It takes us two full days to tile the kitchen, pantry and laundry, and we finish with exactly one tile left over.

The concrete we covered could have been naturally formed, flat as a salt lake, grey as granite. But this floor can't be mistaken. It's man-made, man-laid. You could land a B-36 on it and scrub off the skid marks with a mop. Its grids and angles are a child's great roadwork. You could spot a spider on it at twenty paces. Heather is pleased. A new kitchen floor. We stand out in the middle of this formerly empty room and find it full, wall to wall. We wait only for breakfast.

The remainder of the downstairs, excluding the foyer and the bathroom, we plan to floor in tongue-and-groove oak. Heather and I both like a floor that you can look at for entertainment. A wood floor is that, in almost any composition, because its grain naturally draws the eye. The grain of a long board can lead you all the way down a hall, whether you're going that way or not. A knot in the middle of a wood floor is the epicenter of the world. It's impossible not to explore it. To find two boards cut from the same tree is as astonishing as twins from the womb. A wood floor is, at last, a wider tree limb, something we can trust; bugs may live there, but coyotes don't.

This flooring resembles true Victorian solid oak flooring, but it's really only oak on top, a cedar/oak plywood underneath. The boards are all two inches wide and come in varying lengths. Though they're designed to be glued down, it takes a rubber mallet to set them into place, to lock the tongue into the groove. I begin in the living room, since it's biggest and might least show my learning mistakes. Experience has led me to expect them. The wood-flooring glue is brown, twice as thick as the tile glue, and I'm supposed to use the side of the trowel notched in quarter-inch segments. Pushing the glue across the concrete floor is like a long bout of constipation. The strain is constant. The glue spreads in clean quarter-inch ribbons. It seems a shame to cover it up, mash it between rock and board. But with the first flooring board out of the box I forget all about the glue. These boards are too pretty to put on the floor. I ought to make a case for a fine gun or a writing desk or a jewelry box with this lumber. The grain in some of the pieces is quarter-sawn, or tiger oak as some call it, almost marbled. I could get lost in the grain of this twenty-four-inch board for the rest of the day. But I'm wealthy. I have box after box of such boards, enough to floor two thousand square feet. I lay the first section in the corner of the room, tap it down into the glue with a white rubber mallet. The next board keys into a slot in the first. I like working with the rubber mallet. It's good to really pound something and not hurt it, somewhat like a Bozo Bopper. The lumber seems to be cut perfectly, straight and square, but occasionally a long piece will be bowed and require quite a healthy pounding for tongue to fit into groove. It's a big puzzle where every piece you pick up fits exactly wherever you first put it down. Heather's only aesthetic concern is that I not let the ends of two adjacent boards match up. Since she's handing me the boards, I let her attend to this nicety. She goofs once, but I don't let her in on it. It's my little secret in the floor, the only four-inch horizontal line in the house.

It takes days. We work from living room to dining room to the two downstairs hallways to the downstairs library, building a great flat ship's deck. We bring the dogs out while we work and they sniff the new floors, getting a background scent so they'll better know later on when someone's dropped a steak along the baseboard. They look for throw rugs to lounge on but can't find any and finally choose individual sunny spots. I work in the headiness of my glue, my whole body a rubber mallet, lost in the dizzying galactic grain. Butch lifts his head suddenly, looks out the window and I smile. He whines uneasily, and Heather asks him what's wrong, but he's a dog and can't tell her.

Darwin passes a hot dog up to Quixote and says, "They didn't have mustard."

"Unbelievable," the man of La Mancha says.

"Do you want it or not?" Darwin asks.

Quixote takes the hot dog, his armor clanking as he sits back down, leaning dangerously over the back of the bleacher.

"You're going to fall," Darwin warns.

"You uninitiated," Don Quixote smirks, and then shakes his head sadly.

"I spit on your shoes!" Darwin hisses.

"Sancho Panza!" Quixote screams at him. "Sancho, Sancho, Sancho!"

Darwin puts his hands to his ears and looks down at the littered ground beneath the bleachers; at the very moment that he recalls that he too is a knight of the realm he spies a nickel.

Butch whines again, raises himself and trots to the window to smear dog-nose mucus on it. Heather comforts him, and Butch turns back to her and groans in pleasure. "That's what you saw, isn't it?" she affirms. "Me scratching you." And then looking out, she asks, "Did you see a rabbit?"

Quixote stands on the top bleacher and roars, "A RABBIT?!"

"Heather!" I yell. "I need another rabbit!"

She turns, a smile on her mug, and says, "A rabbit?" and then bursts into gags of laughter.

"A board, I need another piece of flooring," I say.

When Heather and I speak, we have to yell, because whenever Heather works she brings out her boom box. She still enters my life like a squall. She has to try out every new section of flooring with some dance improvisation, explaining that she's making sure it's glued down well. Depending on the music her body is an accumulation of knotted Slinkys or a fluff of down off the wing of a swan floating on a mild breeze. Puppy tries to dance with her, but Butch, as lazy as I am when it comes to unneeded effort, reclines on the new floor and watches. I spread more glue, tacking myself more firmly to this floor, this house, this life, this wife.

In the roar of rock I hear it. I can hardly believe my ears. I rush upstairs, passing Reggie, who's white head to toe, and upstairs again to the third floor and to the back window. Butch is on my heels. We arrive just in time to see the B-36 taxi off the dirt road, snapping limbs off trees with its wings, and turn toward our house. The six propellers and four jets blow the whole farm back behind them. The plane is so huge that the cockpit is at the same level as me, some thirty feet off the ground. Quixote, the idiot, is waving his sword at it. The house shudders, vibrates with the engines. The nose of the plane inches forward and finally squeaks against the window I stand behind. My hand reaches out, palm flat, to know that glass.

I'm anxious. The floor is sweeping through the house. Living room, dining room, hall, hall, library and then upstairs hall and library, the wood goes down faster and faster till we're confronted with the fact of an almost-finished house. Then in one day, the carpet company comes and lays carpet in the three bedrooms and linoleum in the bathrooms. The only unfloored areas in the house, the only cement and plywood left, are the foyer and the entire third floor (which we're leaving till later). Heather and I, after the carpet stretchers have gone, roll on the floor like dogs and cats. We collect carpet fuzzies and hold them out in handfuls to each other. We go barefoot, sloshing through the fronds, and then, strangely enough, thrust our noses into the weave and inhale new carpet smell. Suddenly I realize that Thoreau never did this. I mean, he may have had a throw rug but certainly not wall-to-wall carpeting. I can't decide what this means for me. Am I soft, or just practical? Then even more suddenly, watching Heather scooch and burn on the carpet, I'm thinking all and only of sin. Sin on the new carpet, deep pile comfort, weave and mesh, inhale, girl.

31

Perhaps the Lake Country house never felt like a home simply because I wanted to feel like a visitor there. And that's what I was for the next three years, a visitor, to most places. I lived in five different apartments, a room, a dormitory, a house, a couple of girlfriends' apartments and a series of hotels. The only common characteristic I recall all of these places having was a flat place to lay a writing pad on.

In January of 1980 I loaded my desk into the back of a truck and drove it to Dallas, to a lower-level room and bath of an apartment. Tom and Sylvia lived upstairs, an unmarried couple in their forties who seemed to be more unsure of their own union than they were of me, their boarder. Tom was often out of work and occasionally drunk, occasionally high, and likable in a relaxed, two-day-beard way. Sylvia gushed at having a writer in the house and gave me free roam of the kitchen. They would argue in whispers till Tom couldn't stand the constraint any longer and roared, "Well, I don't care if he can hear us!" and then he'd whisper again. In the morning Sylvia slipped down the stairs and tapped lightly on my door, saying she hoped that I wasn't bothered by all the bother. I said no, trying to make out the outline of her breast beneath her nightgown, the one she'd worn to make up with Tom during the night.

I seldom roamed their kitchen, preferring the toaster oven on my bathroom counter. It cooked a Nighthawk frozen dinner in thirty-five minutes. I kept a bag of Doritos and a bag of Deluxe Grahams under the sink with the toilet plunger, and just down the street on Lemmon there was a 7-Eleven for cold Pepsis. My room had no window but did have a glass patio door guarded by a heavy vinyl curtain. I left the curtain open till I felt the world could see in better than I could see out. This happened in dusk somewhere. I'd stop writing or reading then, draw the curtain and entertain my dinner. I tried to concentrate on everything.

I was an English major now and my course for the semester consisted of five English classes and a writing seminar. I found it hard to believe they'd give me a college degree for reading and writing, but it seemed to be the case. I was boyish toward my professors. I was fond of them before they wrote their names on the board. The first time I left Marshall Terry's office (he was head of the Creative Writing Department) I bumfuzzled his name and said, lightly, "I'm looking forward to everything, Mr. Marshall," as if he were Matt Dillon and I were Festus Hagen. I studied poetry with Laurence Perrine, author of *Sound and Sense*, and fiction with Ken Shields that first spring of 1980. There was such depth and variety in our work, so much joy and honest sadness, that I knew I'd never be able to go back to accounting. I sat in my fiction class, watching Ken Shields stroke his fine grey beard, listening to his voice modulate from sureness to wonder to sureness, and I was stricken with the possibility of a rich life. I say stricken because it's a frightening thing, facing your future, no matter how distinct and wonderful you feel it may be.

I was sitting at my desk on a Saturday morning, pen in hand, when my father called to tell me Zeke had died. I told him I'd come on home. I put my dirty clothes in my dad's old Air Force duffel bag, and then I drove those familiar miles.

Zeke lived with us for twelve years. I don't think he was ever our dog as much as he was canine against the sky, dogdom's dog. He took joy where he could find it, in basic things: food, the chase, sex, a ball. The only prejudice he harbored was against beards; a bearded man could not cross his path, though Zeke never bit a human as far as I know. He handled pain and loneliness better than any other individual I've ever met. I'm sure his offspring still survive, disseminating Zeke's good traits through the species; in California, New Jersey, Indiana, Kentucky, Texas, wherever you run into the big brown eyes of a bird dog, that's Zeke. I became, on that windswept hour's drive home to my dog's funeral, a character of my own imagination. I'd read myself before. A boy and his dog. It's hard to believe at first that any one life is as interesting as another, but I came to have that faith. So I owe a great deal to my dog, that ebony crotch-sniffer, big-pawed hand-shaker. His snout had gone grey, his belly full of old bullets, and he had died.

32

Henry and Emily come in across the field below the house, over the dam of the pond, and sit on the very end of my bleachers. The bleachers are so crowded Emily has to sit in his lap. Emily sits, smooths the skirt of her dress over her thighs and locks her hands around her knees. She looks one way, then another, then glances back at Henry and quickly away and says, in very small letters, shaped like clover, "ahem."

I'm hanging a train light in our kitchen. It's not a blinking railroad track crossing signal, but an old Pullman car fixture converted from kerosene to electricity. It's a heavy, solid brass fixture. Trains didn't need to save weight. After I connect the wiring I fight the weight up to the ceiling, push a two-inch screw into the sheetrock and then drive it through the sheetrock and into a two-by-eight that I, amazingly enough, thought ahead to position. It takes another seven screws to give me the confidence I need to walk under it. I can't imagine how heavy it would be if the two kerosene tanks on the sides of the light were full. I tested the light before installing it, but Heather's been by the switch for ten minutes now, growing fiercely impatient.

"Okay," I say. "Try it."

She flips the ivory switch with her pointed index finger, and we have our, so to speak, first light.

Things are coming together here. The new flooring seems to have put a fire under us as well. I've ordered a thousand board-feet of oak lumber so we can begin to situate our interior doors, cabinets, stairs, and trim. Heather is going over her fiftieth configuration idea for the kitchen cabinets. We've made exploratory trips to appliance stores. We've piled all of our light fixtures, ceiling fans, sinks, toilets, tubs and doors just inside the back door, in the kitchen. The Pullman lamp highlights the jumbled accouterments of our house, the gewgaws, the options.

The light fixtures seem the logical starting point. If we can get them in we'll work in the evenings as well; it's also beginning to become warm at midday, and the ceiling fans could provide some relief. This, however logical, proves to be the incorrect course.

Heather informs me that the correct course is to begin by installing a toilet. She is tired, she informs me, of driving up to my mother's to pee. She is even more tired of my suggestion to find a closer bush.

"A toilet," I explain to her, "is the tip of an iceberg of a water system. I'll have to dig a ditch from the well to the house, and bring the well tank in before I can install a toilet.

"Are you afraid?" she asks. "You're afraid," she says.

"I'm not afraid."

"Well."

I hate digging ditches. I had envisioned a day hanging fine antique brass light fixtures. Instead, I pick up burial tools and begin the grave of a twenty-five-foot piece of copper. The freeze line isn't very deep in Texas, so I only dig down about sixteen inches, till the pick strikes the pumpkin-colored clay. Dad comes down and solders on my copper connections, and I lay my pipe to rest.

We drain the well tank then, watching the forty gallons work their way down the hill to the pond. The tank is shaped something like the atomic bomb Fat Boy and is so difficult to carry that we end up rolling it into the house and pantry. I could leave the tank outside, but I'd have to build another house around it to keep it warm, and I don't think I have the firmness of psyche for that at this point. So I bring the tank into my house like a poor relative, hoping I won't be bothered by the space he requires. While Dad plumbs the tank into the house system, I lay another pipe in the ditch and run an electrical wire through it, connecting the well pump to the breaker box in the pantry.

All of this is the front end of our water system; the rear end, our septic tanks, I know very little about, as they arrived while Heather and I were away for a weekend. All I know is, we had grass one Friday and on Monday we had a central heap of fresh earth with long fingers of fresh earth leading out from it. I assume there are a couple of five-hundred-gallon concrete tanks beneath the surface with perforated pipe leading out from them to dissipate the water into the ground. But as I say, I do not know; it could be the grave of a gigantic tarantula. At any rate a four-inch pipe leaves my house on its way to the earth heap.

Installing a toilet is an ordinary event. It is such a simple job that the directions are printed on the box. Why is it that Heather and I are thrilled at this moment in our lives? She sits on our bathroom floor in a Buddha position watching everything I do, handing me tools, praying for my success. She never does this when I sit down to write one of my books. I bolt the commode to the drain and to the floor and then bolt the tank to the bowl. I screw a valve to the copper pipe coming out of the wall, and connect the valve to the toilet. It's nerve-racking, knowing that water is about to shoot through all of your work. I lean back.

"Ready?" Heather asks.

"Ready," I say.

She bolts downstairs to turn the valve in front of the well tank. I hear the water rushing toward me in gushes and spits, forcing air in front of it through the valve of the toilet. My heart slams against my brain, which is sure that it has forgotten some important washer and that my face will soon receive a fine spray to prove it. The water fires into the tank and soon the sound becomes more familiar, more relaxing, as if my own bladder has just been relieved. Heather yells up at me, and since I can't see any leaks I tell her to come on up. The tank finishes filling while we stand over it. I give Heather the honor. She reaches for the lever gingerly, and presses down with the two joined fingers of the Boy Scout pledge. The bottom drops out, and we watch with pride the fall and counterclockwise rotation of our little Northern Hemisphere whirlpool, our goldfish graveyard, our dog-watering, diaper-dipping, hangover-hugging porcelain friend. Heather quickly becomes absorbed by the lever, flushing again and again, till I have to carry her away kicking and screaming. She's been a switch, button and lever addict for many years. I have to hold her on my lap, my arms wrapped around hers, whenever we get into a friend's power-windowed car.

Our cast-iron, freshly re-porcelained tubs, all three of them, sit in the kitchen at the foot of the stairs. The only person I can find who will help me carry the two big ones upstairs is my little brother, who's still trying to get me to like him. The tubs are unbelievably heavy. We put towels under the rim and still are only able to manage one step at a time. Phil lifts on the front of the tub and pulls, and I lift and push. Every four steps, then every two steps, and finally every step we have to stop and rest. It's like pulling a dead whale up a steep beach. Our only advantage is also our disadvantage, the great inertia of the beast: once you have moved it and set it down again, even on this slope, it doesn't want to move. I appreciate this as I am under it, and don't think I could run down the stairs as fast as it could fall. Once we reach the last stair and pull it up over the bank as it were, quick-stepping it to the bathroom is almost simple. What's not simple is to get back to the stairs, look down and see the second tub still in the kitchen.

The old pedestal sink we're going to put in the master bath comes up with almost as much difficulty, but at least it comes in two pieces. The base is a fluted column, and the sink itself is square with an oval basin. The faucet is nickel-plated with the brass showing through on the handles and the neck of the spout. It's been well-used, and although it's probably seventy-five years old it's the only piece we didn't have to re-porcelain. The rubber washers are worn out, brick-hard and brittle, so I replace them with an amalgam of washers from an assortment kit. It's nice to be able to save the old faucet, because a good new one runs as much as $350. The hardware for our tubs, including a brass and porcelain shower head, totals $650. Somehow we had a mistaken notion that we'd save money by using salvage. The tubs and sinks themselves were reasonable, but the restoration and hardware come at unnegotiable prices. Nostalgia costs. It takes almost two weeks to get all the plumbing fixtures in. I have parts specially made at a machine shop to join the 1911 Kohler faucet to modern plumbing, and the faucet still drips. I change washers half a dozen times, but the faucet still drips. I decide that if I leave the faucet alone the drip will probably fix itself in a few weeks.

The downstairs bath takes a full week in itself. We have to paper the walls before moving in the fixtures because there'd be very little room to work in afterwards. Heather has gone through some seven hundred wallpaper books to find the correct paper for this five-by-six-foot room. It's pink with flowers on it. Somehow I had an idea it would be pink with flowers on it. We've bought all the appropriate wallpapering tools: a long, flimsy plastic tub in which to wet the prepasted paper, smoothing brushes, a squeegee and a little wooden roller for air bubbles and seams. We cut the first sheet of paper quite amiably, rolling it back up and wetting it, "booking" it and actually hanging it on the wall. The problem arrives eight seconds later when the paper falls back off the wall. For some reason, at the exact same moment, Heather and I decide this problem is each other's fault. There follows an excruciating hour wherein she insinuates that I don't even know how to unzip my pants, much less make love to a woman. At the end of the hour we give up, two full sheets of wallpaper crumpled on the floor. Back at the wallpaper store the clerk tells us that, even though prepasted, wallpaper still needs more paste. We buy a bucketful. Heather and I step back into that small ring with our brushes and squeegees and find that a good pot of glue can make a happy marriage.

We move a tiny clawfoot tub in, and an oval pedestal sink. I use all new hardware on these so they go in easily. The commode, our Kellogg mansion waterfall closet, is another story, however. It's English, and none of its female orifices match American plugs. The drain on the bottom of the toilet is too large for a three-inch pipe, so I have to chip the pipe out of the concrete, set a wax ring and drop the commode into that, hoping it won't leak. The water inlet isn't near the size of any standard thread, so I use a radiator hose and two clamps to attach it to the pipe coming from the tank. Luckily this can't be seen from above. The old oak tank above the bowl is American, and all I have to do to it is replace the valves inside. I attach an old cabinet knob to a chain and screw the chain to the lever coming out of the tank, and slowly I turn the valve that releases the water. Water sprays all over the ceiling and runs down our new wallpaper.

"What's happening?" Heather screams.

I shut the valve as quickly as I can, water running down my face and arms. When I climb the ladder and look down into the tank, I see I've left the overflow tube off. I don't tell Heather this. I tell Heather some adjustment is still required. When the tube is on, Heather turns the valve and the tank begins to fill. It doesn't seem to leak. I expected several leaks, what with the old copper tank and the new seals. When the tank is filled I take hold of the cabinet knob on the end of the chain and give it a gentle pull. Nothing happens. We're both gazing at the toilet bowl. I pull a bit harder. The valve releases, water rushes down the five feet of two-inch brass pipe, and Heather and I both jump back through the bathroom door and out into the hall. This commode is a tempestuous, frothing Conradian sea. The breaching foam reaches bidet height. We warn each other never to flush while sitting down, but agree that guests can learn this lesson on their own.

In the meantime our oak has arrived, one thousand board-feet of three-quarter-inch hardwood, random lengths and widths. Some of the boards are eighteen feet long and as much as fifteen inches wide. I don't know if I'll be able to bear cutting them. We immediately take some of this lumber to Frank, a local cabinetmaker. He's already been out to the house to measure the kitchen, and Heather has finally decided on a format for the cabinets. There won't be any upper cabinets, as the kitchen is surrounded by windows. The lower cabinets will consist mainly of deep drawers and deeper drawers. There will even be drawers behind the cabinet doors. Even the trash can will sit in a drawer. She wants everything to come to her.

The day Frank and his wife deliver the cabinets Heather loses control. She jumps up and down and puts all ten of her fingers in her mouth at once. Frank smiles and nods, as if he's seen this behavior dozens of times. I had no idea new kitchen cabinets had this effect upon a woman. Frank and I slide the cabinets into place and screw them to the wall. He uses a sabre saw to cut a hole in the plywood top for the sink. We'll set tile on the plywood. Frank blows all the sawdust off his cabinets, and I pay him for my cabinets. Then I load him up with more oak so he can glue up boards that will be wide enough to use for stair steps. I could use the wide planks I've got, but they'd probably cup with seasonal moisture changes.

Heather is already brushing sanding sealer over her cabinets. She coats and sands with steel wool, then coats again, sands again and finally a third coat of lacquer. I drill for and screw on white porcelain cabinet knobs. She tells me what will go in each drawer and cabinet. "This kitchen will cook by itself," she says. I doubt this but am hesitant to put forth my views. Any subject that will cause a woman to put all ten of her fingers in her mouth shouldn't be bandied about.

The next day we stand on our back porch and hold our hands up when the appliance delivery truck is six inches from the porch roof. If new cabinets make Heather happy, make her revert to schoolgirl fingers in facial orifices, appliances send her back to the womb. I am sure, when the refrigerator is rolled into our house, that she is at the brink of falling on her back and sucking on her toes. She caresses the fingerprint-proof refrigerator door as if it were Bumby, our rabbit. I have already seen every feature of our dishwasher, clothes dryer and washer at the appliance store several times, but Heather takes the time to point them all out to me again. I nod, and instead of looking at the appliance, I watch her. She doesn't notice.

I walk into our house every morning and test the sound characteristics, yelling, "Hallooooo," and then listening not for an answer but for the reverberation and echo of my own voice. Can it be that this is my house? That we're almost done? The work of late has been all attachment, adding jewels to a mount or setting. I move from ceiling fan to light fixture to water heater to kitchen stove at what is to me breathtaking speed, exhilarating moments of accomplishment. Between the living room and foyer I'm finally able to install the huge piece of antique fretwork and its supporting columns. This takes only thirty minutes but casts the shadow of tracery one hundred years old. The bare white walls, backdrop and shadow-catchers, immediately become Victorian walls. The whole structure is to me as some *arc de triomphe*, and a gateway to the future and past. I walk back and forth under its delicate stick and ball and scrollwork, changing the direction of the smooth flow of millenia. Who can tell me what time it is? How has love changed? How much mistletoe has hung from this gingerbread?

Through May and June, in between tubs, toilets, chandeliers and counter tops, I rip oak down to size, run it over a Sears molding blade, inhaling the sweet rotten-fruit stink of steel through oak. Homer, who's handy again, helps me cut frames for our heavy old oak doors and hang them throughout the house. The grain of the new oak isn't as ribald as that of the old (it's quarter-sawn), but the oak is plainly oak and serves the doors well with strength and color. With what's left (and it's plain to see I'll have to buy another thousand feet even to begin on the stairs), Bobby and I trim the baseboards of the first floor, good high eight-inch baseboard that you could slop a mop up against without touching the wall.

During the first week of July, the heat becoming used to itself and our funds depleted again, we realize we're going to have to move in before we finish. We'll have guests from Maine within ten days. Most of the trim, the oak stairways and the black-and-white marble foyer will have to wait, along with minor items such as doorknobs. On the third of July we make one last, sixteen-hour effort. We'll move in tomorrow. Heather sweeps and picks up the entire four thousand square feet, and I install the last of our antique light fixtures in the foyer and hallways. Homer and his wife Debbie help with the cleanup and snap all the screens onto the windows. I'm working on the three-way fixtures, lights you can turn off and on from two different places. For the last time, I curse Mark Laughlin's very existence. After the fixtures are in I find that our, his, perusal of the simplified wiring manual was speed reading at best. One switch will turn off the light but won't turn it on. The other will turn it on but not off. After much re-reading I learn that we left out an entire wire. This wire was supposed to travel through three walls and a ceiling, now completely covered with sheetrock. My father comes down and saves Mark's life (I was loading my shotgun in the car for the trip to Wyoming) by suggesting I simply use the ground wire and not tear out half the walls in my house to put in a new one.

In the evening, after the last dustpan has been emptied and I've packed all my tools behind a closet door, Heather and I walk through this house. "It's big," we say. "It's clean." She brings me a Pepsi from our humming refrigerator. We walk through as we've done so many times, but this is the last time all the same. From now on we'll have to avoid furniture, hold back curtains, turn knobs. This is the last time this house will smell like sawdust and glue. Tomorrow it will probably smell like pizza. The dusk darkens and we almost leave but pause to sit on the tailgate of our truck and look again. I look at our house, at the night sky, and then at my lovely, bone-tired wife. Her T-shirt hangs loosely from her shoulders and gathers in her blue-jeaned lap. Her eyes catch light, and oftentimes I'm sure she's going to cry but she says it's not true.

"All the lights work now?" she asks me.

"Yes, I put the foyer and attic lights and the chandelier in the third-story tower today. That's all of them but a couple closet lamps."

"That won't matter," she says, and jumps off the tailgate and skips into the house.

"What?" I yell. "What?" following her.

She has to grope for the first light switch, but after that she's off, switch to switch to switch, turning on every light in the place. I get the idea and take the old scrap-lumber stairs all the way to the third story and turn on the lights there. We meet back on the second floor and race each other down to the earth. The old Rainbow Bread screen door slams when we're thirty yards into the field below our house. When we're a hundred yards out into the dark night, lost in cockleburs and coastal hay, we turn, winded, afloat in our own breath, to catch sight of that lighthouse that will bring us home.

"If I could be sure of you . . . " Emily says, leaving him an opening, a second chance.

"Take me for granted," Henry says, "for granted. I am a given."

33

Through the next two years I read, wrote more short stories, bought myself a tweed jacket with elbow patches and lived in a studio apartment with greasy walls and bright green shag carpet but with many windows. My desk was beneath one of these windows and equidistant from my bed and the kichen, an excellent tactical position for writing a novel. I reviewed ten or twelve first novels for the *Dallas Times Herald,* and finished mine in the summer of 1982, sending it off to an agent in New York. My mother was the only other person at home when I received the letter a couple of months later. I read it, my heart thumping behind my eyes. I handed the letter to her, saying, "Well, what about that."

She read it, and said, "Well, what about that."

Someone actually wanted to give me money for my book.

My father was recovering from surgery, and having left Pinetree behind he began to look at several businesses and seemed most interested in an old lumberyard twenty miles outside of Fort Worth. I walked over it with him. It began as a salvage yard but now carried new lumber as well. Used railroad crossties dominated the yard. All of the buildings on the property, including the house, were made of materials that had come out of other buildings. The house had no two windows that matched. The corrugated tin roofs on the yard buildings had so many nail holes they resembled the tin panels of an old pie safe. Most of the inventory was split and warped. What little hardware there was in the tiny store was dust-coated and out of date. The only truly viable part of the business was a stake-making shed that provided bundles of forming stakes to concrete crews. Dad and I agreed it would be wonderful work in Texas in August. My father thought the place had promise. I said why not look around some more. When I got on a British Caledonian jet bound for London in February, the last thing my father said to me was, "We might own a lumberyard when you get home."

My parents moved while I was away. The house we now lived in wasn't a mobile home in the technical sense but it had been moved onto the present lot from somewhere else. I thought it strange to imagine the house traveling down the American highway, free from any earthly bondage, trailing its wiring and pipes, its windows seeing the sights. When it arrived, it was married to another house that may or may not have been built on site. The marriage was ill-advised, ill-taken, and spawned several ugly children, outbuildings and garages built of scraps and leftovers, materials that even a salvage lumberyard couldn't sell: warped, half-burned timbers, odd-sized windows, mismatched shingles. The portion of our house that had run the roads had some finer points: wood sashes, oak flooring, a fine rock fireplace. Its mate had linoleum over particle-board flooring, aluminum windows and very dark knotty-pine paneling. My mother made brave efforts with this house. She was promised it was only a temporary abode, so she didn't make a coordinated plan of attack (wallpaper, carpet, curtains). She lived in it, but tried not to touch it, holding her arms tightly to her sides as she walked through its doorways.

It was a two-bedroom house, and as my brother and sister had both moved to their own apartments, I took the second bedroom. The oak floor was covered with that ubiquitous green shag carpet, and the walls were bare waferboard. I pulled up the carpet and painted the walls white, used a putty knife to pry open the long-shut sashes.

As soon as I had the windows open I shut them for good again. The house and lumberyard sat just twenty-five feet or so from State Highway 199, a two-lane road that should have been six. Soon after Dad arrived, an activist had brought a stack of bumper stickers to the hardware store, "Pray for me, I drive Hwy. 199." The road was a constant stream of commuters and gravel trucks, which isn't a good mix for concrete. Later that fall I totaled my TR-3, pulling into our gravel driveway off 199 when a semi tried to pass me over a double yellow line. His front bumper caught my car on the left rear quarter, squeezing the metal up behind my head and spinning the entire car several times in the gravel. Mark Laughlin, who'd come down from Kentucky to work in the hardware store, was sitting next to me. I asked if he was all right, and he replied, "Yes, what happened?" We were both fine, due, I suppose, to someone praying for us.

I worked in the lumberyard a few days each week and worked on a novel a few days each week. Mark worked full time and slept on the floor in my bedroom, and we fell into that self- and buddy-deprecating search for our life's loves. Mark drove a Toyota motorhome, which is a conspicuous vehicle on a first date. We bought safari clothing, vests and hats, and wore them to movies to bolster our confidence in ourselves. We talked to girls in record stores, bookstores, appliance stores, but none of them was amazed that we were available. We found that very few beautiful women walk into a hardware store, and that when they do they've been sent by their husbands.

I became an initiate of cedar and pine, flush valves and pipe wrenches. It was one of the best jobs in the world for a writer. The knowledge I gained at Dad's lumberyard underlies and gives strength to every sentence I write. It was a world of natural resources—lumber, brick, sand, mortar and nails—and a vast variety of parts—plumbing, electrical, and mechanical—and every tool required to put them all together, to build shelter, to provide heat, to keep the human animal occupied with the miracle of his hands. Very few of these things came wrapped in plastic, and so one was able to torture or play with them at will. I spent hours in the plumbing bins, forming inconceivable joints. It's remarkable how much pleasure may be attained by spinning a revolving nail bin. My hands were addicted to the many small boxes of nuts and washers and bolts. But my favorite part of the job, at least when it wasn't 110 degrees outside, was working in the yard, moving lumber and crossties around with our antique forklift, stacking and sorting and sawing sweet lumber. You could walk past the cedar bins and be lifted to a love of the planet that no church could inspire, and I knew because I'd recently been in the best of them. People came to our church with their sadnesses, a leaking toilet, a rusty lock, a broken window, and we solved their problems for an advertised fee. My father often fixed washers and ceiling fans over the phone. He was our high priest.

It didn't take long before Mark and I realized that our small bedroom wasn't big enough for a budding novelist and a budding survivalist. One of the debris-built structures in back of the house was a garage/washroom/storage barn about twenty-five feet by forty feet. My father offered it to us. The building was a shell, but it had a sound roof and concrete floor. We evicted a dozen dead rats, their carcasses as dry and light as crackers, and one torpid but quite poisonous copperhead snake. I designed the apartment mathematically: one third living room, one third bedroom and one third kitchen and bath. I raised the floor under the kitchen and bath and ran plumbing there, which created a conversation pit of the living room. Mark and I put railing around this pit, carpeted it and thought it a fine place to entertain women. I built the kitchen counters and cabinets out of oak pallets, ripping the lumber on our big saw in the yard. I paid twenty dollars for a gas cookstove, and Dad gave us the refrigerator from the hardware store. I paneled the bathroom with cedar and installed two cast-iron wall sinks left by the previous owners. We carpeted the bedroom and raised a portion of its floor to a lofty enough height to support our desks and books, a precipitous decision since we later found that a good rain tended to swamp the bedroom. Home at last. From here I could eat pork chops, write novels and come in as late as I wanted without waking my parents.

The Decatur Road was published in the fall of 1983, and first came to me, a box of ten, while I sat behind the counter in the hardware store. My whole family was there, as we'd been waiting for the arrival of the UPS van for days. I had to fight my father to open the box. After sweeping off a layer of styrofoam peanuts I lifted the top copy out of the box as if it were a Gutenberg bible, holding it with both hands. My parents cooed. I turned away with it then, turned my back to them all, as they descended on the box, a pack of book wolves. I turned to the first page, and read the first paragraph all the way through to make sure it was really mine, and then I let myself smile broadly for many minutes. It's a strange experience, seeing the very words out of one's mouth between hardcovers. The book gave me a queer thrill of overcoming, as if I'd just killed a bear with a bone and knew I'd live through the rest of the day without any more adversaries.

Through the winter and spring, while I worked on my second novel, Dad built a new steel hardware store onto the front of the house. He and Mom had also bought an old farm eight miles further out, 165 acres of brush and grass, and had begun to build a large log home on it. The spruce logs were trucked in from Colorado, and we pinned them together with sledges and twelve-inch nails. By fall my parents had moved out of the house behind the hardware store and into their blue log house with its wrap-around porch, and our family had a homeplace again.

In August of that year I spent three weeks outside of Ranchos de Taos, New Mexico, working for SMU as a creative writing instructor at Fort Burgwin. I had looked forward to the pond as a time of rest, away from the lumberyard, away from a break-up with my girlfriend of two years. My editor's comments on the book I'd just finished would reach me there. There would be fifteen undergraduates at the Fort, but twelve of them were taking Charlie Smith's advanced class. I would be coaching the introductory students, all three of them. So I had good reason to anticipate some quiet time in the mountains, a restful casita in which I could speak softly with my students and work on my manuscript. I made it a point to drive the six hundred August miles across north Texas, the Panhandle, and into New Mexico without turning on the radio in my new little Dodge pickup. There was so much music on that journey: the zip of my tires on the hot asphalt, the wind over my mirror, the moan of all that aridness and the small, tired towns I whispered through. I thought I could write about a becalmed but somehow desperate person living in that flat country.

At Fort Burgwin (now Cantonment Burgwin), the brown needles of piñon clinging to my socks, I stood in the entrance of my casita while Charlie introduced me to the first of his students to arrive. Although she was a junior, she didn't appear to be older than sixteen. The pink blouse and shorts she wore seemed three sizes too large for her. Charlie is no more than six feet tall, but this girl was a full foot shorter than he was. She said, "Hi," timidly, and seemed desperate to be somewhere else, out from under the tall piñons and two male teachers. I looked down at her feet while Charlie talked about the program. She had on hightop sneakers, which startled me because I'd just written an entire book about a girl who wore hightop sneakers. I'd given the trait to the character because I thought it might make her unique. But the girl before me put my character's black basketball hightops to shame. Hers were purple, with day-glo orange shoestrings. She emphasized the shoes by splaying her feet like a duck in repose. Her heels touched but the toes pointed off at almost a full 180 degrees. The boisterous shoes hardly matched the femininity of her pink suit, and neither seemed to coincide with her shyness, the demureness of her manner and what dawned on me as a completely natural beauty. She wore purple tennis shoes but not a hint of makeup. I picked at the needles in my socks, trying to balance on one foot, keep up the flow of conversation and look at her face as often as I could.

That evening eight of us played team Trivial Pursuit in Charlie's cabin. Although she was never my partner, I'd turn to her often and she'd already be looking at me. She had lost all of the shyness of our afternoon meeting and laughed at every word out of my mouth; she laughed at me with only the least provocation. I am somehow predisposed to love anyone who laughs at my jokes, who sees the planet as humorously as I do. I attribute them with intelligence, an appreciation of the arts and a love of dogs. From this point on I deceived, feinted and altered my natural pace that I might sit at her table come breakfast and dinner. On an excursion to the Taos grocery we simultaneously plucked boxes of Apple Jacks cereal from the shelf. Our eyes met. There was a geology class at Fort Burgwin as well; they came to breakfast every morning with tool belts jangling, rock picks, tape measures, hammers, quite professional. Heather and I purchased eggbeaters and wore them on our belts to breakfast next morning, refusing to explain ourselves.

I gave her my copy of *Kentucky Love* to read one afternoon, while I went over the editor's notes I'd just received in the mail. When I got to Chapter Twelve, the first and only sex scene I'd ever written, I blushed severely at the comments, even though I was sitting alone in my casita. My editor advised that I never, never, should use the word "throb" in a sex scene. All I could think about was the beautiful girl now reading my horribly miswritten sex scene, the word "throb" pounding in her skull, convincing her of my sickness as a human being. I had used "throb" three times in a four-page chapter. My only consolation was that I'd used it in three different tenses. I blushed and lost twenty-five years of accumulated self-confidence. But later she brought my manuscript back, dropped it on the chair next to me and left without a word. Her eyes were brimming with tears, and so I let her go.

I still owe her a great deal for that moment. It gave me the courage to ask her out to a movie, and she accepted, and we went in my truck to see *The Natural* at the Taos Drive-in. I told her, as we sat there, that a writer's life was wonderful, because to me, at that moment, it was. She said she thought it must be. She said her grandfather had been a writer all his life, that he'd won two Pulitzer Prizes. She did not elaborate. I looked out at an almost pure baseball player set against a pure desert night and wondered who this girl next to me could be. She didn't look like one of the Hemingway girls. At last I turned to her and said, "Well, what's his name? Faulkner? Hemingway? What?" Another loss of self-confidence. I kissed her a week later and within six months asked her to marry me. Our engagement would last a year and a half. In the meantime she had to finish at SMU.

So I went back to SMU too, not as a student, or even an instructor, but as a boyfriend. I sat in her dorm room while she was in class and wrote in these fifty- and eighty-minute sessions *A Flatland Fable*, the novel I'd first recognized on the way out to Taos. I put 120,000 miles on my pickup, driving between my garage apartment in Azle and the SMU campus. It was lonely at home. Mark had at first left to go back to school at University of Texas at Arlington, and then he'd gotten married to a redhead and moved home to Kentucky. I worked only two days a week at the lumberyard and spent the remainder of my time in Dallas or on the road to and from there. I kept a notepad suction-cupped to my windshield to make best use of all the driving time.

That spring Heather and I went to an antique auction in Grand Prairie. A small English wardrobe came across the auction block, and Heather had to have it for the bedroom in the garage apartment. We bid and bought. Twenty minutes later another wardrobe came across, and I felt for the first time the perplexing and somewhat painful jab of her index finger between my ribs. "That's the one I want." We owned it within thirty seconds. No one in my family wanted our first wardrobe so we stood it in the hardware store with a "For Sale" sign. It sold within two days. Within a month we'd opened an architectural antique shop next door to the lumberyard.

Heather was still in school, so I worked the counter, selling old fireplace mantels, stained-glass windows, doorknobs, anything originally attached to an old house. I had at last become a junk man, an occupation I had desired since the age of six when an old man and his mule came down my grandmother's alley. Two months before we married we opened our first antiques mall, on the far side of the lumberyard. We built a fifteen-thousand-square-foot metal building, attached a small false front, and leased booth space to antique dealers. Within three months the building was full with the varied wares of a hundred dealers. The dealers brought in their merchandise, displayed and priced it, and Heather and I sold it for them. By the end of the year we had another mall in Burleson, south of Fort Worth, some two hundred dealers and eight employees. We'd become a business, but more importantly, I had gained some free time to write and build a house.

On April 5, 1986, Heather and I married on the ground we'd chosen to build our house on. What would be our main hallway was a leaf-blown aisle in our church. Heather wore an antique wedding dress she'd bought for forty dollars, I rented a tuxedo and we stood in front of a double-wedding-ring quilt that my Grandma Coomer had made for us. Bryan Woolley read our vows, a string quartet played under the oaks and buzzards wheeled overhead. I held her face in my hands.

After eight months of marrige we found that although a garage apartment is sufficiently commodious for two young men, it's no more than a thimble for a married couple who collects antiques. Our hallways were like rabbit paths through underbrush. We had no space clear enough to land a plate, and so ate with our food in our laps. By winter we'd retreated to the only territory left to us, our bed. We ate, read, watched TV, did our bookkeeping and slept there. The dogs had worn great ruts in the small yard; not a blade of grass could grow. Our rabbit was tired of his view. Only the cats, writhing in and out of our crammed collection, found life bearable. They were sure mouse families lived inside each new piece we acquired.

We lay in bed late one night, the light out, and decided we had to do something. We had to begin. We knew we didn't have enough money, but our businesses were successful, and we thought we could do some of the work ourselves. We'd consecrated our ground more than a year earlier. My little sister and her husband had already built their house on the farm. I turned to Heather and said, "Tomorrow we'll get graph paper and start."

"Tomorrow we'll start," she said, and she fell asleep easily then, as she always has. She seemed to me to act as if the future were hers, as if vigilance weren't required. I lay there, trying to sleep, trying not to think, shifting my position on the planet, attempting to cache myself in some niche of safety till the morning broke and I could be sure again.

34

Many people older than I am tell me the world existed before I was born, but this is hard for me to believe. It seems to me they have a vested interest in this assertion: first come, first served. The only thing I recall before I was born is a vast nothingness. Perhaps that's why the present has such meaning to me, why I try to draw out eternity, as if it won't last.

I've been showing people through the house lately, a tour I've been happy to guide as we approach our move-in. Last night, our last in the garage apartment, I dreamed, and dreamed I was showing my completed house to my grandfather, my father's father, the carpenter. I haven't seen him in years, since a week before he died. Even in the dream I cherished the moment. He nodded at me as I pointed things out, and said, "Gee-O, Joe, look what you've done," and he smiled often. It was fine to hear his heavy breathing again, to watch him lean with his extended arm against a wall or porch post. I watched his eyes move over my house, blessing it. I watched him sit on my stairs in his overalls and cap and smoke a cigarette. He turned one way, then another, looking for Cerilda, and then he told me once more how proud he was of me, such a good, solid house. Then I woke, my eyes awash in tears, throat hard, my hand almost unbearably empty. I'd been holding a hammer in my dream.

It takes Heather and me an hour to pack the first vanload. It's only an eight-mile drive, so we don't pack too carefully. The last thing we throw in before we roll down the door on the van is Blossom the cat. When we arrive we find she has taken affront at being treated like baggage. She has dumped on a box of books. We skid her into the house and she immediately hits the deck. She mews spastically and crawls along the floor as if barbed wire were strung above her. She runs from room to room mewing, and finally finds a closet dark enough to hide in. She still misses Eliot terribly.

Heather and I carry in furniture and boxes, short-stepping with the heavier pieces to their new rooms and dropping them. There's no time to decorate now; we just want to get it all in the door. I realize as we move from room to room, during the back and forth, that my bleachers have emptied. They've all come inside. I left the door wide open. They mill, walk from room to room, and I can't help but smile. Thoreau is going through my books. Quixote is asleep on the couch. It has taken me twenty-nine and a half years to build the house that I live in. The house I live in tomorrow will take me twenty-nine and a half years and one day to build. Eliot, our kitty, out of his long sleep, jumps up in my wife's arms. And Crazy Mike, of my Refugio childhood, asks if he might stay in the closet under the stairs. I ask him sternly if he will be kind to my children, and he nods, looking down at the pebbles in his palm. I realize he needs a home too. Ever since I've started building this house, writing this book, I've continually called Butch "Zeke." He doesn't seem to mind. I recognize them all, my characters, my grandfather, my heroes, all but the little boy in the horn-rimmed glasses, but then I see the baseball glove on his hand, the stack of books under the other arm, and I realize that I've become a fictional character myself. Van load after van load, our house begins to fill, to look and smell and breathe like us. After we've loaded the last load out of our garage apartment, we stand in that emptiness, that past, and feel it gone, feel sadness and memory and happiness oozing in, filling the void. On the way out to the house I say, "We're going home."

I feel now that I, too, am experienced in adventures.

I think if Thoreau surveyed me now he'd find me about twenty percent desperation. But with each page I write I become more, or less, desperate. The more I try to define my faith the more I cheapen it somehow, make it less. So I simply say I have faith and let it go at that. I bring my parents and grandparents into our home, showing them where we'll put our Christmas tree. I want to tell them there's no need to worry about us now. I want to tell them we're safe.

What will last? Of Veal Station, only a marble memorial, a tombstone. Of the Tonkawa, shots that missed the bird or buck and were lost in the grass. And before them? Only the life-born rock left by a receding ocean long, long ago. And of our house, what will last? Certainly not this stick-and-nail Victorian structure. It's not even complete yet, we haven't lived in it one full day, and I already see it grey, weathered, burning, Captain Veal's ghost standing in the doorway. Perhaps in fifty years I'll bring my grandchildren to this bramble-thickened hillside and complain that someone's let all this good country grow up. For a while the foundation will stand. We built on a hillside so our concrete piers had to go down a full eight feet for support. When the house is gone and all the soil beneath it has eroded to fill the pond below, the slab will be left high in the air, held up by our twenty-five piers: a Stonehenge for the future perhaps, somehow found to align with the moon and stars. I take some solace in this.

Solace too in a survivor rose we've found, next to a sunken spot in old Veal Station, blooming a mottled pink every spring. It must be a hundred by now. Solace in a few old chimney stones we reused in our rock wall; solace in this printed page, which may outlive our house. And solace, my wife and I, in our realization of what will last: the meantime, our lives together building a home, a family's life to live in it.

I think the only act more optimistic than building a house is having children. People ask us what's through the stained-glass door in our stairwell, and we tell them heaven. I wrote this book for me, but give it to Heather and what children may come. It's dusk now and there's much still to be done, but we're all in. And now, finally, now, our home is tall and cool and silent, and Heather and I are sureness and hope, and have become lately bedridden, down among down, house of joy, body of joy, whisper our names.